FATIGUE DESIGN OF STEEL AND COMPOSITE STRUCTURES

2ND EDITION

ECCS Eurocode Design Manuals

ECCS Editorial Board
Luís Simões da Silva (ECCS)
António Lamas (Portugal)
Jean-Pierre Jaspart (Belgium)
Reidar Bjorhovde (USA)
Ulrike Kuhlmann (Germany)

Design of Steel Structures – 2ND Edition
Luís Simões da Silva, Rui Simões and Helena Gervásio

Fire Design of Steel Structures – 2ND Edition
Jean-Marc Franssen and Paulo Vila Real

Design of Plated Structures
Darko Beg, Ulrike Kuhlmann, Laurence Davaine and Benjamin Braun

Fatigue Design of Steel and Composite Structures – 2ND Edition
Alain Nussbaumer, Luís Borges and Laurence Davaine

Design of Cold-formed Steel Structures
Dan Dubina, Viorel Ungureanu and Raffaele Landolfo

Design of Joints in Steel and Composite Structures
Jean-Pierre Jaspart and Klaus Weynand

Design of Steel Structures for Buildings in Seismic Areas
Raffaele Landolfo, Federico Mazzolani, Dan Dubina, Luís Simões da Silva and Mario d'Aniello

ECCS – SCI Eurocode Design Manuals

Design of Steel Structures, UK Edition
Luís Simões da Silva, Rui Simões and Helena Gervásio
Adapted to UK by Graham Couchman

Design of Joints in Steel Structures, UK Edition
Jean-Pierre Jaspart and Klaus Weynand
Adapted to UK by Graham Couchman and Ana M. Girão Coelho

ECCS Eurocode Design Manuals – Brazilian Editions

Dimensionamento de Estruturas de Aço
Luís Simões da Silva, Rui Simões, Helena Gervásio, Pedro Vellasco and Luciano Lima

Information and ordering details

For price, availability, and ordering visit our website **www.steelconstruct.com**.
For more information about books and journals visit **www.ernst-und-sohn.de**

FATIGUE DESIGN OF STEEL AND COMPOSITE STRUCTURES

Eurocode 3: Design of Steel Structures
Part 1-9 – Fatigue
Eurocode 4: Design of Composite Steel and Concrete Structures

2nd Edition

Alain Nussbaumer
Luís Borges
Laurence Davaine

Fatigue Design of Steel and Composite Structures

2nd Edition, 2018

Published by:
ECCS – European Convention for Constructional Steelwork
publications@steelconstruct.com
www.steelconstruct.com

Sales:
Wilhelm Ernst & Sohn Verlag für Architektur und technische Wissenschaften GmbH & Co. KG, Berlin

All rights reserved. No parts of this publication may be reproduced, stored in a retrieval system, or transmitted in any form or by any means, electronic, mechanical, photocopying, recording or otherwise, without the prior permission of the copyright owner.

ECCS assumes no liability with respect to the use for any application of the material and information contained in this publication.

Copyright © 2018 ECCS – European Convention for Constructional Steelwork

ISBN (ECCS): 978-92-9147-139-3
ISBN (Ernst & Sohn): 978-3-433-03220-6

Legal dep.: 437674/18 Printed in Sersilito, Empresa Gráfica Lda, Maia, Portugal
Photo cover credits: Alain Nussbaumer (front cover), Thierry Delémont (back cover)

TABLE OF CONTENTS

FOREWORD	xi
PREFACE	xiii
ACKNOWLEDGMENTS	xv
SYMBOLOGY	xvii
TERMINOLOGY	xxi

Chapter 1

INTRODUCTION	1
1.1 Basis of fatigue design in steel structures	1
1.1.1 General	1
1.1.2 Main parameters influencing fatigue life	3
1.1.3 Expression of fatigue strength	7
1.1.4 Variable amplitude and cycle counting	10
1.1.5 Damage accumulation	13
1.2 Damage equivalent factor concept	15
1.3 Codes of Practice	18
1.3.1 Introduction	18
1.3.2 Eurocodes 3 and 4	18
1.3.3 Eurocode 9	21
1.3.4 Execution (EN 1090-2)	23
1.3.5 Other execution standards	29
1.4 Description of the structures used in the worked examples	30
1.4.1 Introduction	30
1.4.2 Steel and concrete composite road bridge (worked example 1)	31

TABLE OF CONTENTS

1.4.3 Chimney (worked example 2)	34
1.4.4 Crane supporting structures (worked example 3)	39

Chapter 2
APPLICATION RANGE AND LIMITATIONS — 43

2.1 Introduction	43
2.2 Materials	44
2.3 Corrosion	44
2.4 Temperature	45
2.5 Loading rate	47
2.6 Limiting stress ranges	47

Chapter 3
DETERMINATION OF STRESSES AND STRESS RANGES — 51

3.1 Fatigue loads	51
3.1.1 Introduction	51
3.1.2 Road bridges	52
3.1.3 Railway bridges	57
3.1.4 Crane supporting structures	59
3.1.5 Masts, towers, and chimneys	61
3.1.6 Silos and tanks	70
3.1.7 Tensile cable structures, tension components	70
3.1.8 Other structures	71
3.2 Damage equivalent factors	72
3.2.1 Concept	72
3.2.2 Critical influence line length	75
3.2.3 Road bridges	76
3.2.4 Railway bridges	82
3.2.5 Crane supporting structures	84
3.2.6 Towers, masts and chimneys	92

3.3 Calculation of stresses	93
3.3.1 Introduction	93
3.3.2 Relevant nominal stresses	94
3.3.3 Stresses in bolted joints	96
3.3.4 Stresses in welds	96
3.3.5 Nominal stresses in steel and concrete composite bridges	99
3.3.6 Nominal stresses in tubular structures (frames and trusses)	100
3.4 Modified nominal stresses and concentration factors	104
3.4.1 Generalities	104
3.4.2 Misalignments	107
3.5 Geometric stresses (Structural stress at the hot spot)	113
3.5.1 Introduction	113
3.5.2 Determination using FEM modelling	115
3.5.3 Determination using formulas	117
3.6 Stresses in orthotropic decks	119
3.7 Calculation of stress ranges	122
3.7.1 Introduction	122
3.7.2 Stress range in non-welded details	123
3.7.3 Stress ranges in bolted joints	125
3.7.4 Stress range in welds	131
3.7.5 Multiaxial stress range cases	133
3.7.6 Stress ranges in steel and concrete composite structures	137
3.7.7 Stress ranges in connection devices from steel and concrete composite structures	142
3.8 Modified Nominal stress ranges	146
3.9 Geometric stress ranges	148

Chapter 4
FATIGUE STRENGTH   157

4.1 Introduction	157

TABLE OF CONTENTS

4.1.1 Set of fatigue strength curves	157
4.1.2 Modified fatigue strength curves	162
4.1.3 Size effects on fatigue strength	163
4.1.4 Mean stress influence	165
4.1.5 Post-weld improvements	165
4.2 Fatigue detail tables	166
4.2.1 Introduction	166
4.2.2 Non-welded details classification (EN 1993-1-9, Table 8.1)	166
4.2.3 Welded plated details classification (general comments)	168
4.2.4 Longitudinal welds, (built-up sections, EN1993-1-9 Table 8.2), including longitudinal butt welds	169
4.2.5 Transverse butt welds (EN1993-1-9 Table 8.3)	170
4.2.6 Welded attachments and stiffeners (EN 1993-1-9 Table 8.4), and load-carrying welded joints (EN 1993-1-9 Table 8.5)	171
4.2.7 Welded tubular details classification (EN 1993-1-9 Tables 8.6 and 8.7)	174
4.2.8 Orthotropic deck details classification (EN 1993-1-9 Tables 8.8 and 8.9)	175
4.2.9 Crane girder details (EN 1993-1-9 Table 8.10)	176
4.2.10 Tension components details (EN 1993-1-11)	176
4.2.11 Geometric stress categories (EN 1993-1-9, Annex B, Table B.1)	179
4.2.12 Particular case of web breathing, plate slenderness limitations	180
4.3 Determination of fatigue strength or life by testing	180

Chapter 5
RELIABILITY AND VERIFICATION 183

5.1 Generalities	183
5.2 Strategies	185
5.2.1 Safe life	185
5.2.2 Damage tolerant	185

5.3 Partial factors . . . 186
    5.3.1 Introduction . . . 186
    5.3.2 Action effects partial factor . . . 187
    5.3.3 Strength partial factor . . . 188
5.4 Verification . . . 192
    5.4.1 Introduction . . . 192
    5.4.2 Verification using the fatigue limit . . . 193
    5.4.3 Verification using damage equivalent factors . . . 201
    5.4.4 Verification using damage accumulation method . . . 207
    5.4.5 Verification of tension components . . . 209
    5.4.6 Verification using damage accumulation in case of two or more cranes . . . 210
    5.4.7 Verification under multiaxial stress ranges . . . 212

Chapter 6
BRITTLE FRACTURE . . . 221

6.1 Introduction . . . 221
6.2 Steel quality . . . 223
6.3 Relationship between different fracture toughness test results . . . 224
6.4 Fracture concept in EN 1993-1-10 . . . 229
    6.4.1 Method for toughness verification . . . 229
    6.4.2 Method for safety verification . . . 231
    6.4.3 Flaw size design value . . . 234
    6.4.4 Design value of the action effect stresses . . . 236
6.5 Standardisation of choice of material: maximum allowable thicknesses . . . 238

REFERENCES . . . 247

ANNEX A   STANDARDS FOR STEEL CONSTRUCTION . . . 257

ANNEX B   FATIGUE DETAIL TABLES WITH COMMENTARY . . . 263

B.1 Plain members and mechanically fastened joints (EN 1993-1-9, Table 8.1)   264

B.2 Welded built-up sections (EN 1993-1-9, Table 8.2)   267

B.3 Transverse butt welds (EN 1993-1-9, Table 8.3)   269

B.4 Attachments and stiffeners (EN 1993-1-9, Table 8.4)   272

B.5 Load carrying welded joints (EN 1993-1-9, Table 8.5)   274

B.6 Hollow sections (T ≤ 12.5 mm) (EN 1993-1-9, Table 8.6)   277

B.7 Lattice girder node joints (EN 1993-1-9, Table 8.7)   279

B.8 Orthotropic decks - closed stringers (EN 1993-1-9, Table 8.8)   281

B.9 Orthotropic decks - open stringers (EN 1993-1-9, Table 8.9)   283

B.10 Top flange to web junction of runway beams (EN 1993-1-9, Table 8.10)   284

B.11 Detail categories for use with geometric (hot spot) stress method (EN 1993-1-9, Table B1)   286

B.12 Tension components   288

B.13 Review of orthotropic decks details and structural analysis   290

ANNEX C   MAXIMUM PERMISSIBLE THICKNESSES TABLES   295

C.1 Maximum permissible values of element thickness t in mm (EN 1993-1-10, Table 2.1)   295

C.2 Maximum permissible values of element thickness t in mm (EN 1993-1-12, Table 4)   296

FOREWORD

Steel structures have been built worldwide for more than 120 years. For the majority of this time, fatigue and fracture used to be unknown or neglected limit states, with the exception in some particular and "obvious" cases. Nevertheless, originally unexpected but still encountered fatigue and fracture problems and resulting growing awareness about such have that attitude reappraised. The consequent appearance of the first ECCS recommendations on fatigue design in 1985 changed radically the spirit. The document served as a basis for the fatigue parts in the first edition of Eurocodes 3 and 4. Subsequent use of the latter and new findings led to improvements resulting in the actual edition of the standards, the first to be part of a true all-European set of construction design standards.

As with any other prescriptive use of technical knowledge, the preparation of the fatigue parts of Eurocodes 3 and 4 was long and based on the then available information. Naturally, since the publication of the standards, have evolved not only structural materials but also joint techniques, structural analysis procedures and their precision, measurement techniques, etc., each of these revealing new, previsouly unknown hazardous situation that might lead to fatigue failure. The result is that even the most actual standards remain somewhat unclear (but not necessarily unsafe!) in certain areas and cover some others not sufficiently well or not at all. Similar reasoning can be applied for the fracture parts of Eurocode 3, too.

Having all the above-mentioned in mind, the preparation of this manual was intended with the aim of filling in some of the previously revealed gaps by clarifying certain topics and extending or adding some others. For the accomplishment of that task, the manual benefited from a years-long experience of its authors and its proofreaders in the fields treated in it; it is a complete document with detailed explanations about how to deal with fatigue and fracture when using Eurocodes… but also offering much, much more. This is probably the most exhaustive present-day fatigue manual on

Foreword

the use of Eurocodes 3 and 4, checked and approved by members of ECCS TC6 "Fatigue and Fracture".

This document outlines all the secrets of fatigue and fracture verifications in a logical, readable and extended (in comparison to the standards) way, backed by three thoroughly analysed worked examples. I am convinced that a manual as such cannot only help an inexperienced user in the need of some clarifications but can also be hailed even by the most demanding fatigue experts.

Mladen Lukić
CTICM, Research Manager
ECCS TC6 Chairman

PREFACE

This book addresses the specific subject of fatigue, a subject not familiar to many engineers, but relevant for achieving a satisfactory design of numerous steel and composite steel-concrete structures. Since fatigue and fracture cannot be separated, they are indeed two aspects of the same behaviour, this book also addresses the problem of brittle fracture and its avoidance following the rules in EN 1993-1-10.

According to the objectives of the ECCS Eurocode Design Manuals, this book aims at providing design guidance on the use of the Eurocodes for practicing engineers. It provides a mix of "light" theoretical background, explanation of the code prescriptions and detailed design examples. It contains all the necessary information for the fatigue design of steel structures according to the general rules given in Eurocode 3, part 1-9 and the parts on fatigue linked with specific structure types.

Fatigue design is a relatively recent code requirement. The effects of repetitive loading on steel structures such as bridges or towers have been extensively studied since the 1960s. This work, as well as lessons learned from the poor performance of some structures, has led to a better understanding of fatigue behaviour. This knowledge has been implemented in international recommendations, national and international specifications and codes since the 1970s. At European level, the ECCS recommendations (ECCS publication N° 43 from 1985) contained the first unified fatigue rules, followed then by the development of the structural Eurocodes. Today, fatigue design rules are present in many different Eurocode parts: EN 1991-2, EN 1993-1-9, EN 1993-1-11, EN 1993-2, EN 1993-3, etc. as will be seen throughout this book.

Chapter 1 introduces general aspects of fatigue, the main parameters influencing fatigue life, damage and the structures used in the worked examples. The design examples are chosen from typical structures that need to be designed against fatigue: i) a steel and concrete composite bridge which is also used in the ECCS design manual on EN 1993-1-5 (plate buckling), ii) a steel chimney and iii) a crane supporting structure. Chapter 2

PREFACE

summarizes the application range of the Eurocode and its limitations in fatigue design. Chapters 3 to 5 are the core of this book, explaining the determination of the parts involved in a fatigue verification namely: applied stress range, fatigue strength of details, fatigue design strategies and partial factors, damage equivalent factors. For each of the parts a theoretical background is given, followed by explanation of the code prescriptions and then by application to the different design examples. Finally, chapter 6 deals with steel selection, which in fact is the first step in the design process but is separated from fatigue design in the Eurocodes. In this chapter, the theory and application of EN 1993-1-10 regarding the selection of steel for fracture toughness are discussed. Note that the selection of material regarding through-thickness properties is not within the scope of this book. The books also includes annexes containing the fatigue tables from EN 1993-1-9, as well as detail categories given in other Eurocode parts (cables). The tables include the corrections and modifications from the corrigendum issued by CEN on April 1st, 2009 (changes are highlighted with a grey background). These tables also contain an additional column with supplementary explanations and help for the engineer to classify properly fatigue details and compute correctly the stress range needed for the verification. The last annex contains the tables from EN 1993-1-10 and EN 1993-12 giving the maximum permissible values of elements thickness to avoid brittle fracture.

Luís Borges
Laurence Davaine
Alain Nussbaumer

ACKNOWLEDGMENTS

This document was written under the supervision of the ECCS Editorial Committee. It was reviewed by the members of this committee, whom the authors would like to thank:

> Luís Simões da Silva (Chairman - ECCS),
> António Lamas (Portugal)
> Jean-Pierre Jaspart (Belgium)
> Reidar Bjorhovde (USA)
> Ulrike Kuhlmann (Germany)

The document was also reviewed by the ECCS Technical Committee 6, working group C. Their comments and suggestions were of great help to improve the quality of the document. Many thanks to all contributive former and current members:

Ömer Bucak, Matthias Euler, Hans-Peter Günther (Chairman WG-C), Senta Haldimann-Sturm, Rosi Helmerich, Stefan Herion, Henk Kolstein, Bertram Kühn, Mladen Lukic (Chairman TC6), Johan Maljaars and Joël Raoul.

Many thanks are also due to all the other persons, too numerous to mention here, who offered their continuous encouragement and suggestions. A large part of the figures were made or adapted by ICOM's talented draftsman and more, Claudio Leonardi.

Finally, thanks are due to Ms. Joana Albuquerque for formatting the text before publication.

Luís Borges
Laurence Davaine
Alain Nussbaumer

SYMBOLOGY

This list of symbols follows the Eurocodes, in particular EN 1993-1-9, and only the fatigue relevant symbols are given below.

Latin letters

A	Area
a	Crack depth
b_{eff}	Relevant thickness in Wallin toughness correlation
c	Half crack length
C	Constant representing the influence of the construction detail in fatigue strength expression
m	Fatigue curve slope coefficient
D, d	Damage sum, damage
G	Permanent actions effects
k_f	Stress concentration factor (i.e. geometric stress concentration factor, thus in this publication there is no difference with k_t)
K_{mat}	Fracture toughness
I	inertia
I_2	inertia of the cracked composite cross section
M	Bending moment
N, n	Number of cycles, number
N_{tot}	Total number of cycles in a spectrum
n_0	short term modular ratio, E_a / E_{cm}
n_{insp}	Total number of inspections during services life
n_{stud}	number of shear studs per unit length
P_f	Failure probability
Q	Load
Q_E	Damage equivalent fatigue load
$Q_{E,2}$	Damage equivalent fatigue load related to 2 million cycles
$Q_{K,1}$	Characteristic value of dominant variable load,
$Q_{K,i}$	Characteristic value of accompanying variable loads,
Q_i, Q_{fat}	Characteristic fatigue load

Symbology

R	Stress ratio, $\sigma_{min}/\sigma_{max}$
S	Standard deviation, characteristic value of the effects of the concrete shrinkage
t	Time, thickness
t_0	Reference thickness, equal to 1 mm
T	Temperature
T_k	Characteristic value of the effects of the thermal gradient
T_{KV27}	Temperature at which the minimum energy is not less than 27 J in a CVN impact test
T_{K100}	Temperature at which the fracture toughness is not less than 100 MPa·m$^{1/2}$
$T_{min,d}$	Lowest air temperature with a specified return period, see EN 1991-1-5
ΔT_r	Temperature shift from radiation losses of the structural member
ΔT_σ	Temperature shift for the influence of shape and dimensions of the member, imperfection from crack, and stress σ_{Ed}
ΔT_R	Temperature shift corresponding to additive safety element
$\Delta T_{\dot{\varepsilon}}$	Temperature shift for the influence of strain rate
$\Delta T_{\varepsilon pl}$	Temperature shift from from cold forming

Greek Symbols

γ_{Ff}	Partial factor for fatigue action effects
γ_{Mf}	Partial factor for fatigue strength
λ	Damage equivalent factor
λ_1	Factor accounting for the span length (in relation with the length of the influence line)
λ_2	Factor accounting for a different traffic volume than given
λ_3	Factor accounting for a different design working life of the structure than given
λ_4	Factor accounting for the influence of more than one load on the structural member,
λ_{max}	Maximum damage equivalent factor value, taking into account the fatigue limit.
λ_v	Damage equivalent factor for the connection
ψ_1	Combination factor for frequent loads
$\psi_{2,i}$	Combination factor for quasi-permanent loads

σ_{min}	Minimum direct or normal stress value (with sign), expressed in N/mm^2
σ_{max}	Maximum direct or normal stress value (with sign), expressed in N/mm^2
σ_{res}	Residual stress value, expressed in N/mm^2
v_2	distance from the neutral axis to the relevant fibre in a steel concrete beam
$\Delta\sigma_C$	Fatigue strength under direct stress range at 2 million cycles, expressed in N/mm^2
$\Delta\tau_C$	Fatigue strength under shear stress range at 2 million cycles, expressed in N/mm^2
$\Delta\sigma_D$	Constant amplitude fatigue limit (CAFL) under direct stress range, at 5 million cycles in the set of fatigue strength curves, expressed in N/mm^2
$\Delta\sigma_{E,2}$	Equivalent direct stress range, computed at 2 million cycles, expressed in N/mm^2
$\Delta\sigma_L$	Cut-off limit under direct stress range, at 100 million cycles in the set of fatigue strength curves, expressed in N/mm^2
$\Delta\tau_L$	Cut-off limit under shear stress range, at 100 million cycles in the set of fatigue strength curves, expressed in N/mm^2
Δv_L	longitudinal shear force per unit length at the steel-concrete interface

TERMINOLOGY

Associated Eurocode	Eurocode parts that describe the principles and application rules for the different types of structures with the exception of buildings (bridges, towers, masts, chimneys, crane supporting structures, tanks…).
Classification method	Fatigue verification method where fatigue resistance is expressed in terms of fatigue strength curves for standard classified details. Can refer to both the **nominal stress method** or the **modified nominal stress method**.
Constant amplitude fatigue limit (CAFL)	The limiting direct or shear stress range value below which no fatigue damage will occur in tests under constant amplitude stress conditions. Under variable amplitude conditions all stress ranges have to be below this limit for no fatigue damage to occur.
Constructional detail	A structural member or **structural detail** containing a structural discontinuity (e.g. a weld) for which the nominal stress method is applied. The Eurocodes contain classification tables, with **classified constructional details** and their corresponding detail categories (i.e. fatigue strength curves).
Control	Operation occurring at every important, identified, step during the fabrication process and during which various checks are made (e.g. tolerances control, NDE controls of welds, of paint layer thickness, etc.).
Crack	A sharp flaw or imperfection for which the crack tip radius is close to zero.
Crack initiation life	Crack nucleation time, micro-cracking stage. The portion of fatigue life consumed before a true crack (in the order of magnitude of one-tenth of a millimeter) is produced.

Terminology

Crack propagation life	Portion of fatigue life between crack initiation and failure (according to conventional failure criterion or actual member rupture).
Cut-off limit	Limit below which stress ranges of the design spectrum do not contribute to the calculated cumulative damage.
Cyclic plasticity	Material subjected to cyclic loading up to yield stress in tension and in compression during each cycle. Alternative term for describing oligo-cyclic fatigue.
Design working life	Value of duration of use, lifetime, of a structure fixed at the design stage, also referred to as **design service life**.
Detail category	Classification of structural members and details (i.e. classified structural details) according to their fatigue strength. The designation of every detail category corresponds to its fatigue strength at two million cycles, $\Delta\sigma_C$.
Direct stress	Stress which tends to change the volume of the material. In fatigue, relevant stress in the parent material, acting on the detail, together with the **shear stress**. In EN 1993-1-9, the above is differentiated from the **normal stress**, which is defined in a weld.
Flaw	Also referred to as **imperfection**. An unintentional stress concentrator, e.g. rolling flaw, slag inclusions, porosity, undercut, lack of penetration, etc. Can be within the production/fabrication tolerances (imperfection) or outside them (defect). In this document, it is assumed that flaws are within tolerances.
Generic Eurocode	Eurocode parts that describe the generic principles for all structures and application rules for buildings (EN 199x-1-y).
Geometric stress	Also known as **structural stress**. Value of stress on the surface of a structural detail, which takes into account membrane stresses, bending stress components and all stress concentrations due to structural discontinuities, but ignoring any local notch effect due to small discontinuities such as weld toe geometry, flaws, cracks, etc. (see sub-chapters 3.5 and 3.9).

Geometric stress method	Fatigue verification method where fatigue resistance is expressed in terms of fatigue strength curves for reference weld configurations applicable to geometric stresses. Also referred to as hot spot stress method.
Hot spot	A point in the structure subjected to repeated cycling loading, where a fatigue crack is expected to initiate due to a combination of stress concentrators. The **structural stress at the hot spot** is the value of geometric stress at the weld toe used in fatigue verification. Its definition, and the related design fatigue curve, is not unique since different extrapolation methods exist.
Imperfection	See flaw.
Inspection	Operation occurring, usually at prescribed intervals, on a structure in service and during which the structure and its members are inspected visually and using NDT methods to report any degradation (e.g. hits and bends, corrosion, cracks, etc.).
Longitudinal	In the direction of the main force in the structure or detail (Figure 0.1).

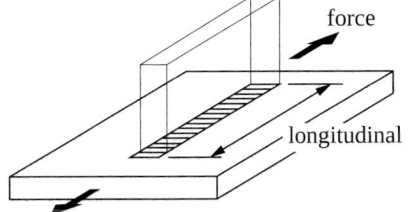 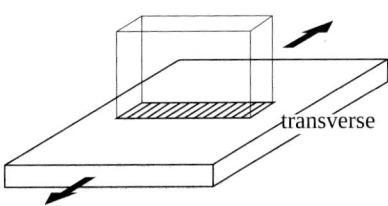

Figure 0.1 – Orientation of the attachment with respect to the main force

Maintenance	Operation made on a structure in service and consisting in corrections and minor repairs on the structure (e.g. painting, cleaning, etc.).
Mean stress	The average between the minimum and maximum stress, i.e. $(\sigma_{min} + \sigma_{max})/2$.

Terminology

Modified nominal stress	Nominal stress increased by an appropriate stress concentration factor to include the effect of an additional structural discontinuity that has not been taken into account in the classification of a particular detail such as misalignment, hole, cope, cut-out, etc. (see sub-chapter 3.4 and section 3.7.7). The appropriate stress concentration factor is labelled k_f or k_1 (for hollow sections joints).
Monitoring	Operation occurring on a structure in service, during which measurements or observations are made to check the structure's behavior (e.g. deflection, crack length, strain, etc.).
Nominal stress	Stress in a structural member near the structural detail, obtained using simple elastic strength of material theory, i.e. beam theory. Influence of shear lag, or effective widths of sections shall be taken into account. Stress concentrators and residual stresses effects are excluded (see section 3.3.2)
Normal stress	A stress component perpendicular to the sectional surface. In fatigue, relevant stress component in a weld, together with **shear stress** components.
S-N curve	Also known as **fatigue strength curve** or **Wöhler's curve**. A quantitative curve expressing fatigue failure as a function of stress range and number of stress cycles.
Shear stress	A stress component which tends to deform the material without changing its volume. In fatigue, relevant stress(es) in the parent material together with the **direct stress** or, in a weld, with the **normal stress**.
Stress range	Also known as **stress difference**. Algebraic difference between the two extremes of a particular stress cycle (can be a direct, normal or shear stress) derived from a stress history.
Stress concentration factor	The ratio of the concentrated stress to the nominal stress (see sub-chapter 3.4), used usually only for direct stresses.

Structural stress Synonym for geometric stress.

Transverse Also referred to as **lateral**. Direction perpendicular to the direction of main force in the structure or detail (Figure 0.1).

Chapter 1

INTRODUCTION

1.1 BASIS OF FATIGUE DESIGN IN STEEL STRUCTURES

1.1.1 General

Fatigue is, with corrosion and wear, one of the main causes of damage in metallic members. Fatigue may occur when a member is subjected to repeated cyclic loadings (due to action of fluctuating stress, according to the terminology used in the EN 1993-1-9) (TGC 10, 2006). The fatigue phenomenon shows itself in the form of cracks developing at particular locations in the structure. These cracks can appear in diverse types of structures such as: planes, boats, bridges, frames (of automobiles, locomotives or rail cars), cranes, overhead cranes, machines parts, turbines, reactors vessels, canal lock doors, offshore platforms, transmission towers, pylons, masts and chimneys. Generally speaking, structures subjected to repeated cyclic loadings can undergo progressive damage which shows itself by the propagation of cracks. This damage is called *fatigue* and is represented by a loss of resistance with time.

Fatigue cracking rarely occurs in the base material remotely from any constructional detail, from machining detail, from welds or from connections. Even if the static resistance of the connection is superior to that of the assembled members, the connection or joint remains the critical place from the point of view of fatigue.

Figure 1.1 shows schematically the example of a steel and concrete composite road bridge subjected to traffic loading. Every crossing vehicle results in cyclic actions and thus stresses in the structure. The stresses

1. INTRODUCTION

induced are affected by the presence of attachments, such as those connecting the cross girders to the main girders. At the ends of attachments, particularly at the toes of the welds which connect them with the rest of the structure, stress concentrations occur due to the geometrical changes from the presence of attachments. The very same spots also show discontinuities resulting from the welding process.

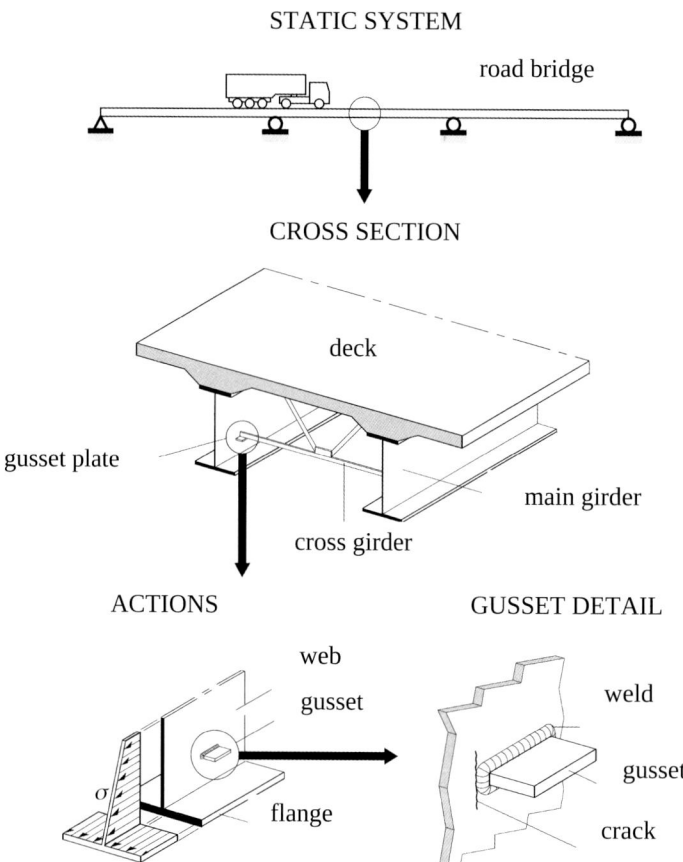

Figure 1.1 – Possible location of a fatigue crack in a road bridge (TGC 10, 2006)

Numerous studies were made in the field of fatigue, starting with Wöhler (1860) on rail car axles some 150 years ago. These demonstrated that the combined effect of discontinuities and stress concentrations could be the origin of the formation and the propagation of a fatigue crack, even if the applied stresses remain significantly below the material yield stress (by applied stresses, it is meant the stresses calculated with an elastic structural analysis considering the possible stress concentrations or residual stresses). A crack develops generally from discontinuities having a depth of the order of some tenth of millimetre. The propagation of such a crack can lead to failure by yielding of the net section or by brittle fracture, mainly depending upon material characteristics, geometry of the member, temperature and loading strain rate of the section. Thus, a structure subjected to repeated cyclic loadings has to be done by careful design and fabrication of the structural members as well as of the structural details, so as to avoid a fatigue failure. The methods of quality assurance have to guarantee that the number and the dimensions of the existing discontinuities stay within the tolerance limits. The purpose of this sub-chapter is to present an outline of the fatigue phenomenon, in order to provide the basic knowledge for the fatigue design of bolted and welded steel structures. To reach this objective, the sub-chapter is structured in the following way:

- Section 1.1.2: The main factors influencing fatigue life are described.
- Section 1.1.3: Fatigue testing and the expression of fatigue strength are explained.
- Section 1.1.4: Variable amplitude and cycle counting.
- Section 1.1.5: Concept of cumulative damage due to random stresses variations.

The principles of fatigue design of steel structures are given in Eurocode 3, part 1-9. For aluminium structures, the principles are to be found in Eurocode 9, part 1-3, fatigue design of aluminium structures. The principles are the same, or very similar, for the different materials. All these standards are based on the recommendations of the European Convention for Constructional Steelwork (ECCS/CECM/EKS) for steel (ECCS, 1985) and for aluminium (ECCS, 1992).

1.1.2 Main parameters influencing fatigue life

The fatigue life of a member or of a structural detail subjected to repeated cyclic loadings is defined as the number of stress cycles it can stand before failure.

1. INTRODUCTION

Depending upon the member or structural detail geometry, its fabrication or the material used, four main parameters can influence the fatigue strength (or resistance, both used in EN 1993-1-9):

- the stress difference, or as most often called *stress range*,
- the structural detail geometry,
- the material characteristics,
- the environment.

Stress range

Figure 1.2 shows the evolution of stress as a function of the time t for a constant amplitude loading, varying between σ_{min} and σ_{max}. The fatigue tests (see following section) have shown that the *stress range* $\Delta\sigma$ (or stress difference by opposition to stress amplitude which is half this value) is the main parameter influencing the fatigue life of welded details. The stress range is defined by equation (1.1) below:

$$\Delta\sigma = \sigma_{max} - \sigma_{min} \qquad (1.1)$$

where

σ_{max} Maximum stress value (with sign)
σ_{min} Minimum stress value (with sign)

Other parameters such as the minimum stress σ_{min}, maximum stress σ_{max}, their mean stress $\sigma_m = (\sigma_{min} + \sigma_{max})/2$, or their ratio $R = \sigma_{min}/\sigma_{max}$ and the cycle frequency can usually be neglected in design, particularly in the case of welded structures.

One could think, a priori, that fatigue life can be increased when part of the stress cycle is in compression. This is however not the case for welded members, because of the residual stresses (σ_{res} in tension introduced by welding). The behaviour of a crack is in fact influenced by the summation of the applied and the residual stresses (see Figure 1.2). A longer fatigue life can however be obtained in particular cases, by introducing compressive residual stresses through the application of weld improvement methods, or post-weld treatments, after welding (see section 4.1.5).

1.1 BASIS OF FATIGUE DESIGN IN STEEL STRUCTURES

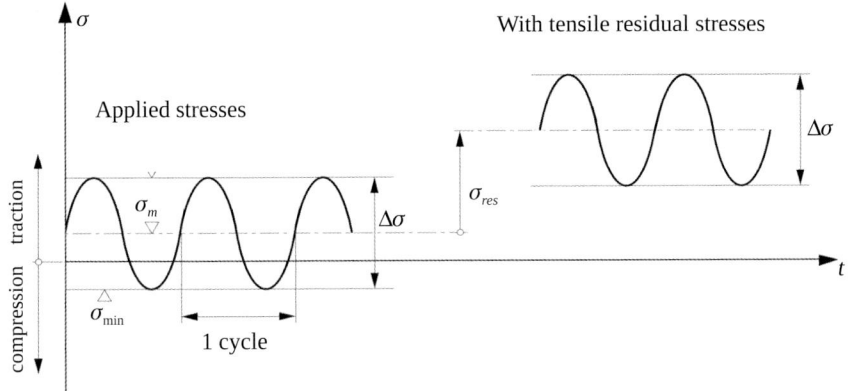

Figure 1.2 – Definition of stresses and influence of tensile residual stresses
(TGC 10, 2006)

Structural detail geometry

The geometry of the structural detail is decisive in the location of the fatigue crack as well as for its propagation rate; thus it influences the detail fatigue life expectancy directly. The elements represented in Figure 1.1 allow to illustrate the three categories of geometrical influences:

- effect of the structure's geometry, for example the type of cross section;
- effect of stress concentration, due to the attachment for example;
- effect of discontinuities in the welds.

The effects of the structure's geometry and of the stress concentrations can be favourably influenced by a good design of the structural details. A good design is effectively of highest importance, as sharp geometrical changes (due for example to the attachment) affect the stress flow. This can be compared to the water speed in a river, which is influenced by the width of the river bed or by obstacles in it. In an analogous manner, stresses at the weld toe of an attachment are higher than the applied stresses. This explains why stress concentrations are created by attachments such as gussets, bolt holes, welds or also simply by a section change. The influence of discontinuities in the welds can be avoided by

1. INTRODUCTION

using adequate methods of fabrication and control, in order to guarantee that these discontinuities do not exceed the limiting values of the corresponding quality class chosen using EN 1090-2 (see section 1.3.4 for detailed information). Besides, it must be clarified that discontinuities in the welds can be due to the welding process (cracks, bonding imperfections, lack of fusion or penetration, undercuts, porosities, etc.) as well as to notches due to the rolling process, or to grinding, or also to corrosion pits. According to their shape and their dimension, these discontinuities can drastically reduce the fatigue life expectancy of a welded member. The fatigue life can be further reduced if the poor detail is located in a stress concentration zone.

Material characteristics

During fatigue tests on plain metallic specimens (i.e. non-welded specimens) made out of steel or aluminium alloys, it has been observed that the chemical composition, the mechanical characteristics as well as the microstructure of the metal often have a significant influence on the fatigue life. Thus, a higher tensile strength of a metal can allow for a longer fatigue life under the same stress range, due essentially to an increase in the crack initiation phase and not to an increase in the crack propagation phase. This beneficial influence is not present, unfortunately, in welded members and structures, as their fatigue lives is mainly driven by the crack propagation phase. In fatigue design, the influence of the tensile strength of the material has usually been neglected; there are only a few exceptions to this rule (machined joints and post-weld treated joints in particular). As a rule of thumb, the fatigue resistance of constructional details in aluminium can be taken as $1/3$ of those in steel, which is the ratio between the elasticity modulus of the materials.

Environment influence

A corrosive (air, water, acids, etc.) or humid environment can drastically reduce the fatigue life of metallic members because it increases the crack propagation rate, especially in the case of aluminium members. On one hand, specific corrosion protection (special painting systems, cathodic protection, etc.) is necessary in certain conditions, such as those found in offshore platforms or near chemical plants. On the other hand, in the case of weathering steels used in civil engineering, the superficial corrosion occurring in welded structures stay practically without influence on the fatigue life expectancy; the small corrosion pits responsible for a possible fatigue crack initiation are indeed less critical than the discontinuities normally introduced by welding.

The influence of temperature on fatigue crack propagation can be neglected, at least in the normal temperature range, but must be accounted for in applications such as gas turbines or airplane engines where high temperatures are seen. A low temperature can, however, reduce the critical crack size significantly, i.e. size of the crack at failure, and cause a premature brittle fracture of the member, but it does not affect significantly the material fatigue properties (Schijve, 2001).

Finally, in the case of nuclear power stations, where stainless steels are used, it is known that neutron irradiation induce steel embrittlement (English, 2007), thus making them more prone to brittle failure (chapter 6) and also reducing their fatigue strength properties.

1.1.3 Expression of fatigue strength

In order to know the fatigue strength of a given connection, it is necessary to carry out an experimental investigation during which test specimens are subjected to repeated cyclic loading, the simplest being a sinusoidal stress range (see Figure 1.2). The test specimen must be big enough in order to properly represent the structural detail and its surroundings as well as the corresponding residual stress field. The design of the experimental program must also include a sufficient amount of test specimens in order to properly measure the results scatter. Even under identical test conditions the number of cycles to failure will not be the same for apparently identical test specimens. This is because there are always small differences in the parameters which can influence the fatigue life (tolerances, misalignments, discontinuities, etc.). The test results on welded specimens are usually drawn on a graph with the number of cycles N to failure on the abscissa (or to a predefined size of the fatigue crack) and with the stress range $\Delta\sigma$ on the ordinate (Figure 1.3).

1. INTRODUCTION

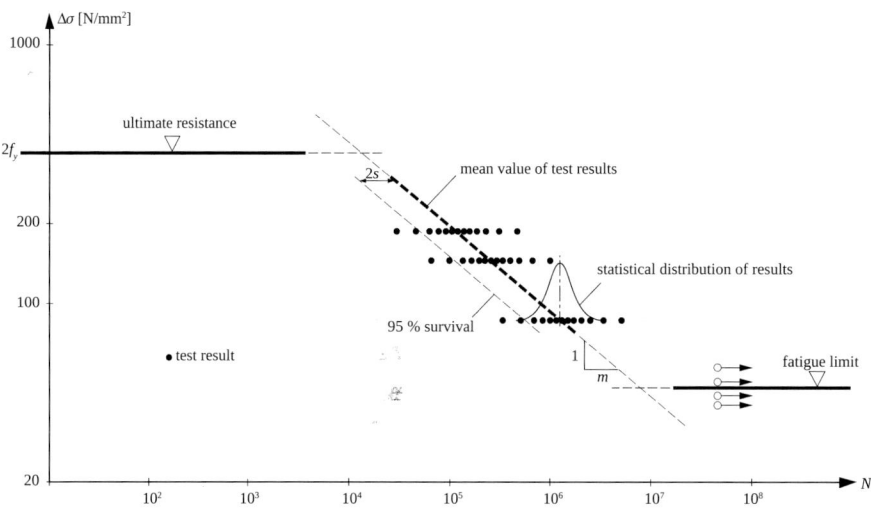

Figure 1.3 – Fatigue test results of structural steel members, plotted in double logarithm scale, carried out under constant amplitude loading (TGC 10, 2006)

The fact is that the scatter of the test results is less at high ranges and larger at low stress ranges, see for example Schijve (2001). By using a logarithmic scale for both axes, the mean value of the test results for a given structural detail can be expressed, in the range between 10^4 cycles and $5 \cdot 10^6$ to 10^7 cycles, by a straight line with the following expression:

$$N = C \cdot \Delta\sigma^{-m} \tag{1.2}$$

where

        N number of cycles of stress range $\Delta\sigma$,
        C constant representing the influence of the structural detail,
        $\Delta\sigma$ constant amplitude stress range,
        m slope coefficient of the mean test results line.

The expression represents a straight line when using logarithmic scales:

$$\log N = \log C - m \cdot \log(\Delta\sigma) \tag{1.3}$$

The expressions (1.2) and (1.3) can also be analytically deduced using fracture mechanics considerations (TGC 10, 2006).

The upper limit of the line (corresponding to high $\Delta\sigma$ values) corresponds to twice the ultimate static strength of the material (reverse cyclic loading). The region with number of cycles ranging between 10 and 10^4 is called low-cycle fatigue (or oligo-cyclic fatigue, with large cyclic plastic deformations). The corresponding low-cycle fatigue strength is only relevant in the case of loadings such as those occurring during earthquakes, or possibly silos, where usually members experience only small numbers of stress cycles of high magnitude.

The lower limit of the line (corresponding to low $\Delta\sigma$ values) represents the constant amplitude fatigue limit (CAFL, or also endurance limit). This limit indicates that cyclic loading with ranges under this limit can be applied a very large number of times ($> 10^8$) without resulting in a fatigue failure. It explains the wider band scatter observed near the fatigue limit, which results from specimens that do not fail after a large number of load cycles (so-called run-out, see Figure 1.3). This value is very important for all members subjected to large numbers of stress cycles of small amplitude, such as those occurring in machinery parts or from vibration effects. One shall mention that investigations for mechanical engineering applications have shown that at very high number of cycles, over 10^8 cycles, a further decline of the fatigue resistance of steels exists (Bathias and Paris 2005). Also, for aluminium, no real fatigue limit can be seen, but rather a line with a very shallow slope (with a large value of the slope coefficient m). It is also important to insist on the fact that a fatigue limit can only be established with tests under constant amplitude loadings. In order to derive a fatigue strength curve for design, i.e. a characteristic curve, the scatter of the test results must be taken into account. To this goal, a given survival probability limit must be set. In EN 1993-1-9, the characteristic curve is chosen to represent a one-sided 95% tolerance bound of survival probability, with a 75% confidence (e.g. a confidence interval on the mean equal to 75%). The exact position of the strength curve also depends upon the number of the available test results. This influence may be accounted for using the recommendations published by the International Institute of Welding (IIS/IIW) (IIW, 2009).

For a sufficiently large number of data points (in the order of 60 test results), this survival probability can be approximated by a straight line parallel to the mean line of the test results, but located on its left, at a two standard deviation $2s$ distance (see Figure 1.3).

1. INTRODUCTION

1.1.4 Variable amplitude and cycle counting

Remember that the curves used to determine the fatigue strength, or S-N curves (e.g. Figure 1.3), were determined with tests under constant amplitude loadings (constant $\Delta\sigma$ stress ranges) only. However, real loading data on a structural member (for example, as a result of a truck crossing a bridge, see Figure 1.1), consist of several different stress ranges $\Delta\sigma_i$ (variable amplitude loading history). Therefore, it raises the question of how to count stress cycles and how to consider the influence of the different stress magnitudes on fatigue life. To illustrate the subject, Figure 1.4 gives an illustration of a generic variable stress history.

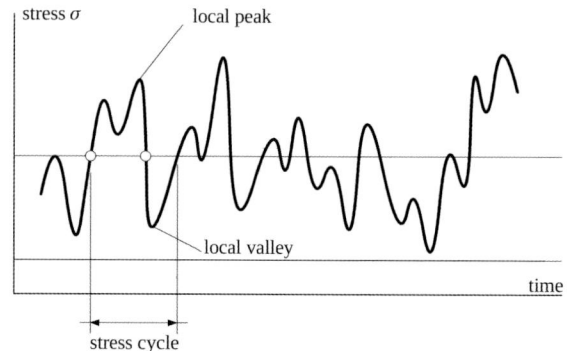

Figure 1.4 – Illustration of generic variable amplitude stress-time history (ECCS, 2000)

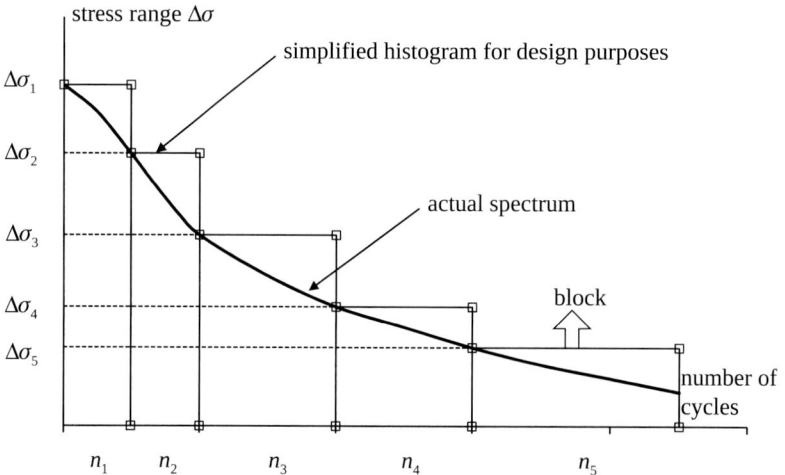

Figure 1.5 – Example of stress spectrum and corresponding histogram (ECCS, 2000)

There are various methods allowing for an analysis of the stress history: peak count methods, level crossing count methods, the rainflow count method and the reservoir count method. Among these methods, the last two shall preferably be used. These two methods, which give identical results if correctly applied, allow for a good definition of the stress ranges; this is of the highest importance since, as seen in section 1.1.2, it is the main parameter influencing fatigue life (i.e. with respect to other parameters such as the maximum or mean stress values). As a result, any stress history can be translated into a stress range spectrum. The algorithm for the rainflow counting method can be found in any reference book on fatigue, as for example Schijve (2001) and IIW (2009). An example of spectrum is shown in Figure 1.5. The spectrum can be further reduced to a histogram, any convenient number of stress intervals can be chosen, but each block of stress cycles should be assumed, conservatively, to experience the maximum stress range in that block histogram.

The rainflow counting method has found some support in considering cyclic plasticity. Also, some indirect information about sequences is retained because of the counting condition in the method, in opposition to level crossing or range counting methods (i.e. if a small load variation occurs between larger peak values, both the larger range as well as the smaller range will be considered in the Rainflow counting method) (Schijve, 2001). it is also this method that is generally suggested to give the better statistical reduction of a load time history defined by successive numbers of peaks and valleys (troughs) if compared to the level crossing and the range counting methods. Two main reasons for preferring rainflow counting (Schijve, 2001):

1) an improved handling of small intermediate ranges
2) an improved coupling of larger maxima and lower minima compared to range counts

The rainflow counting method is thus the best method, irrespective of the type of spectra (steep or flat, narrow-band or broad-band). The other counting methods give more importance to the number and the values of the extrema (peak count), or to the number of crossings of a given stress value (level crossing count). Those are not well suited for welded metallic structures in civil engineering, because of a lower correspondence with the dominating fatigue strength parameters.

As an example (Schumacher and Blanc, 1999), an extract of the stress history measured in the main girder of a road bridge is given in Figure 1.6,

1. INTRODUCTION

and the corresponding stress range histogram after rainflow analysis (corresponding to a total of 2 weeks of traffic measurements) is given in Figure 1.7. The passage of each truck on the bridge can be identified, with in addition a lot of small cycles due to the passage of light vehicles. All cycles below 1 N/mm^2 have been suppressed from the analysis; there is still after the rainflow analysis a significant number of small cycles that can be considered not relevant for fatigue damage analysis (i.e. they are below the cut-off limit, see terminology, of any detail category).

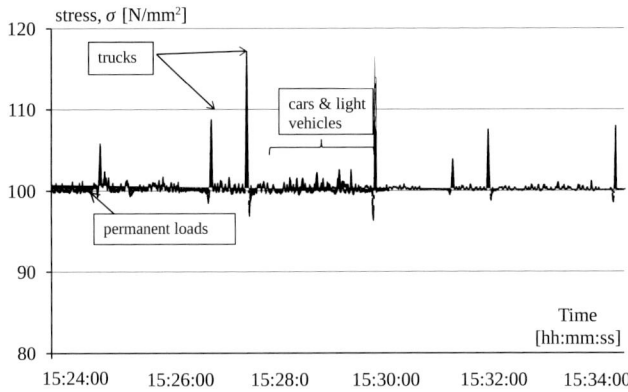

Figure 1.6 – Example of measured stress history on a road bridge (Schumacher and Blanc, 1999)

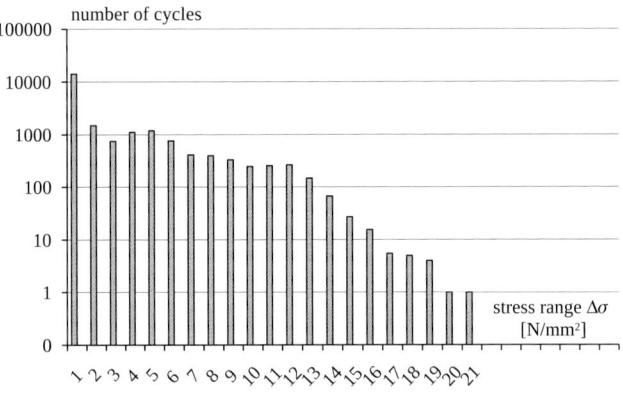

Figure 1.7 – Example of stress range histogram from two weeks measurements on a road bridge (Schumacher and Blanc, 1999)

With help of assumptions on damage accumulation, as explained in the next section, the influence of the various stress ranges on fatigue life can be interpreted with respect to the constant amplitude strength curves

(S-N curves), allowing for the calculation of the fatigue life under real, variable amplitude loading.

1.1.5 Damage accumulation

The assumption of a linear damage accumulation results in the simplest rule, the Palmgren-Miner's rule (Palmgren, 1923)(Miner, 1945), more generally known as the *Miner's rule*. This linear damage accumulation scheme assumes that, when looking at a loading with different stress ranges, each stress range $\Delta\sigma_i$, occurring n_i times, results in a partial damage which can be represented by the ratio n_i/N_i (the histogram being distributed among n_{tot} stress range classes). Here, N_i represents the number of cycles to failure (fatigue life of the structural detail under study) under the stress range $\Delta\sigma_i$. In the case the stress range distribution function is known, the summation of the partial damages due to each stress range level can be replaced by an integral function. The failure is defined with respect to the summation of the partial damages and occurs when the theoretical value $D_{tot} = 1.0$ is reached, see equation (1.4). This is represented in a graphical way in Figure 1.8.

$$D_{tot} = \frac{n_1}{N_1} + \frac{n_2}{N_2} + \frac{n_3}{N_3} + \ldots\ldots = \sum_{i=1}^{n_{tot}} \frac{n_i}{N_i} = \int \frac{dn}{N} \leq 1.0 \qquad (1.4)$$

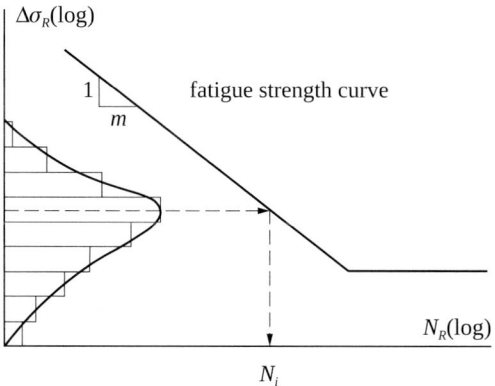

Figure 1.8 – Damage accumulation scheme

It should be noted that in this simple damage accumulation rule, the order of occurrence of the stress ranges in the history is completely ignored, it is thus a simplification. The use in design of equation (1.4) together with suitable safety factors showed itself reliable enough to be considered as the only rule for the fatigue design of welded members of bridges and cranes supporting runways. One shall however be very careful with its applicability to other structure types, especially those subjected to occasional overloads (loads significantly higher than the service loads) such as can be the case in mechanical engineering applications, offshore platforms or in airplanes (IIW, 2009), see section 5.4.4 for more information. All the same, mean stress effect need not be considered when dealing with welded members; they can however be of importance when designing or verifying members of bolted or riveted structures subjected to repeated cyclic loadings, as they can result in significantly longer fatigue lives (compared to the case of every cycle being fully effective in terms of damage as it is the case in welded members).

Stress ranges below the fatigue limit may or may not be accounted for. The first and conservative approach is to ignore the fatigue limit and to extend the straight line with the slope coefficient m.

The second approach takes into account the fact that the stress ranges $\Delta\sigma_i$, lower than the fatigue limit, correspond theoretically to an infinite fatigue life. However, one must be careful because this observation was made under constant amplitude fatigue tests. Applying this rule to variable amplitude loadings only holds true in the case where *all* the stress ranges in the histogram are below the fatigue limit. In this particular case, and only in this one, fatigue life tending to infinity ($> 10^8$ cycles) can be obtained. This is important for given members in machinery or vehicles which must sustain very large numbers of cycles. Let's now look at an histogram with some stress ranges, $\Delta\sigma_i$, above the constant amplitude fatigue limit $\Delta\sigma_D$ as well as others below (Figure 1.9).

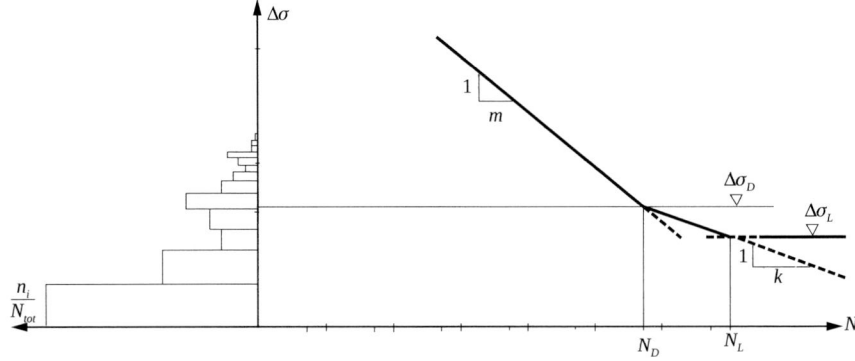

Figure 1.9 – Influence of stress ranges below the constant amplitude fatigue limit, $\Delta\sigma_D$, and the cut-off limit, $\Delta\sigma_L$ (TGC 10, 2006)

In the case the stress ranges are higher than the fatigue limit, the damage accumulation can be computed using equation (1.4). If the stress ranges are lower than the fatigue limit, they do not contribute to the propagation of the crack until the crack reaches a certain size. This is the reason why the part of the histogram below the fatigue limit cannot be completely ignored; it contributes to the accumulation of damage when the crack becomes large. To avoid having to calculate the crack growth rate using fracture mechanics for stress ranges $\Delta\sigma_i$ lower than the fatigue limit a resistance curve is used with a slope k different from Wöhler's slope m ($k = 2m - 1$ according to Haibach (1970) or $k = m + 2$, both giving for $m = 3$ the same value, $k = 5$).

In addition, in order to take into account the fact that the smallest values of stress ranges $\Delta\sigma_i$ do not contribute to crack propagation, a cut-off limit is introduced, $\Delta\sigma_L$. In many applications, including bridges, all the stress ranges lower than the cut-off limit can be neglected for the damage accumulation calculation. The cut-off limit is often fixed at 10^8 cycles, giving $\Delta\sigma_L \approx 0.55 \cdot \Delta\sigma_D$ in the case N_D is equal to $5 \cdot 10^6$ cycles (slope $k = 5$).

It is important to repeat that the part of the fatigue resistance curve (Figure 1.9) below the fatigue limit is the result of a simplification and does not directly represent a physical behaviour. This simplification was adopted in order to facilitate the calculation of the damage accumulation, using the same hypothesis as for stress ranges above the fatigue limit.

For aluminium, the fatigue strength curves follow the same principles as explained above, including the values of the number of cycles, N_D and N_L at which slope changes occur. The only exception is that different values for the Wöhler's slope m were found. These values also differ for structural detail groups. The value of the slope up to the constant amplitude fatigue limit (CAFL) can take the following values: $m = 3.4, 4.0, 4.3$ and 7.0. As for steel, for stress ranges below the CAFL, a strength curve with a slope $k = m + 2$ is used.

1.2 DAMAGE EQUIVALENT FACTOR CONCEPT

The fatigue check of a new structure subjected to a load history is complex and requires the knowledge of the loads the structure will be subjected to during its entire life. Assumption about this loading can be made, still leaving the engineer with the work of doing damage accumulation calculations. The concept of the fatigue damage equivalent factor was proposed to eliminate this

1. INTRODUCTION

tedious work and put the burden of it on the code developers. The computation of the usual cases is made once for all. The concept of the damage equivalent factor is described in Figure 1.10, where $\gamma_{Ff} Q_k$ is replaced by Q_{fat} for simplicity. On the left side of the figure, a fatigue check using real traffic is described. On the right side, a simplified model is used. The damage equivalent factor λ links both calculations in order to have damage equivalence.

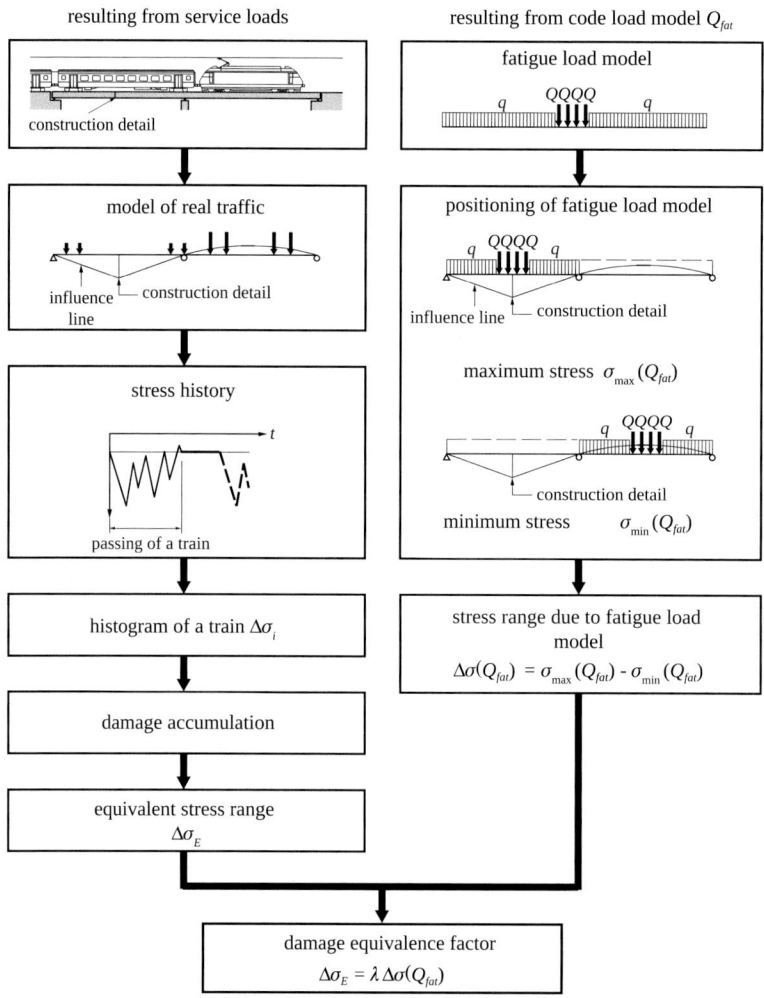

Figure 1.10 – Damage equivalent factor (Hirt, 2006)

The description of the procedure on the left side, procedure that was used by the code developers, is:

3) Modelling of real traffic and displacement over the structure,
4) Deduction of the corresponding stress history (at the detail to be checked),
5) Calculation of the resulting stress range histogram $\Delta\sigma_i$,
6) Computation of the resulting equivalent stress range $\Delta\sigma_e$ (or $\Delta\sigma_{E,2}$ for the value brought back at 2 million cycles), making use of an accumulation rule, usually a linear one such as *Miner's rule*.

Finally, one can perform the verification by comparing $\Delta\sigma_{E,2}$ with the detail category or, if there is more than one detail to check, the detail categories. Note that one can also perform the verification directly by performing a damage accumulation calculation and check that the total damage remains inferior to one (in this case, the detail category must be known beforehand to perform the calculation). Detailed information on damage equivalent factors can be found in sub-chapter 3.2.

This procedure is relatively complex in comparison with usual static calculations where simplified load models are used. It is however possible to simplify the fatigue check, using a load model specific for the fatigue check, in order to obtain a maximum stress σ_{max} and minimum stress σ_{min}, by placing this load model each time in the most unfavourable position according to the influence line of the static system of the structure. But the resulting stress range $\Delta\sigma(\gamma_{Ff} Q_k)$, due to the load model, does not represent the fatigue effect on the bridge due to real traffic loading! In order to have a value corresponding to the equivalent stress range $\Delta\sigma_{E,2}$, the value $\Delta\sigma(\gamma_{Ff} Q_k)$ must be corrected with what is called a damage equivalent factor, λ, computed as

$$\lambda = \frac{\gamma_{Ff}\Delta\sigma_{E,2}}{\Delta\sigma(\gamma_{Ff} Q_k)} \quad (1.5)$$

The calculations of the correction factor values are made once for all for the usual cases, and are a function of several parameters such as the real traffic loads (in terms of vehicle geometry, load intensities and quantity) and influence line length, to mention the most important ones.

The main assumptions are the use of the "rainflow" counting method and a linear damage accumulation rule. Therefore, one may ignore phenomena such as crack retardation, influence of loading sequence, etc. The S-N curves must belong to a set of curves with slope changes at the same number of cycles, but the curves can have more than one slope. This is the case for the set of curves in

1. INTRODUCTION

ECCS, EN 1993-1-9, or for aluminium EN 1999-1-3. The simplified load model should not be too far from reality (average truck or train), otherwise there will be some abrupt changes in the damage equivalent factor values when the influence line length value approaches the axle spacing. The fatigue load models for different types of structures can be found in the various parts of Eurocode 1, see sub-chapter 3.1 for further details. The damage equivalent factor has been further split into several partial damage equivalent factors, see sub-chapter 3.2.

1.3 CODES OF PRACTICE

1.3.1 Introduction

In structural engineering, a great deal of research during the 1960s and 70s focussed on the effects of repetitive loading on steel structures such as bridges or towers. This work, as well as the lessons learned from the poor performance of some structures, led to a better understanding of fatigue behaviour. Still, it was a problem long overlooked in civil engineering codes, but considered in other industries (e.g. mechanical engineering, aeronautical engineering), each industry having its theory and calculation method. The work done in the 60s and 70s in turn led to the first fatigue design recommendations for steel structures and to substantial changes in provisions of steel structures design specifications. The first codes in Europe that considered fatigue were the german code (DIN 15018, 1974) and the British code (BS 5400-10, 1980). It was followed by the first European ECCS recommendations in the 80s (ECCS, 1985), which contained the first unified rules with a standardized set of S-N curves, which is still in use today.

1.3.2 Eurocodes 3 and 4

In Europe, the construction market and its services is regulated through product standards, testing codes and design codes, the whole forming an international standard family. The European standard family prepared by the European Standardization body, i.e. "Comité Européen de Normalisation" (CEN), includes so far 10 Eurocodes with design rules, for a total of 58 parts, and many hundreds of EN-standards for products and testing. It also contains so far around 170 European Technical Approvals (ETA) and European Technical

1.3 CODES OF PRACTICE

Approval Guidelines (ETAG), all prepared by the European Organisation for Technical Approvals (EOTA). For steel structures, the relevant parts of the European international standard family are shown on Figure 1.11.

Apart from the general rules, Eurocode 3 contains "Application rules" like part 2 "Steel bridges" or part 6 "Crane supporting structures" on special ranges of application.

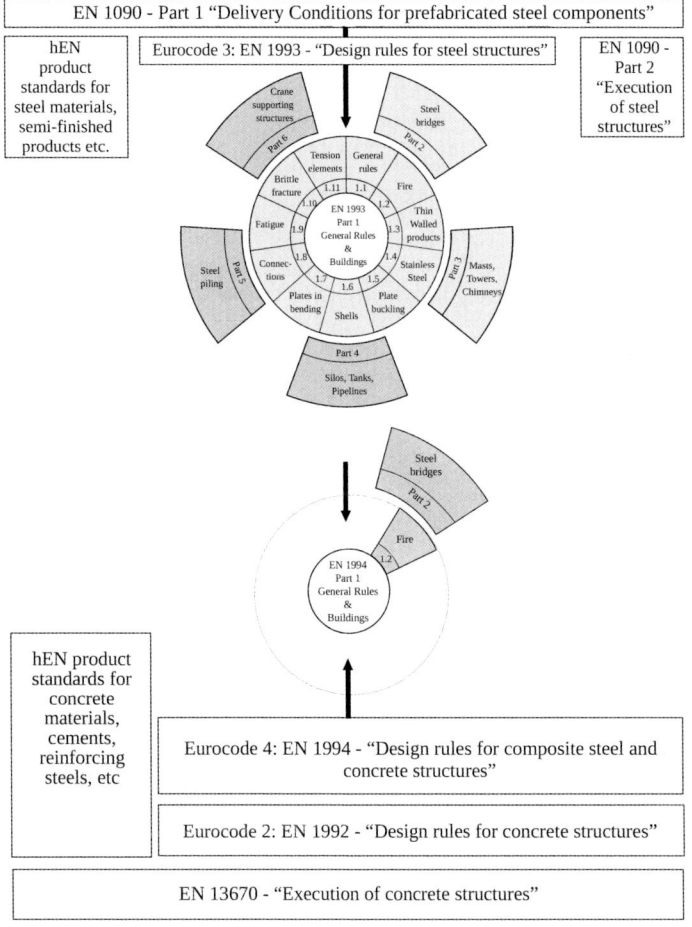

Figure 1.11 – Standard system for steel structures (Schmackpfeffer *et al*, 2005) and composite steel and concrete structures

Altogether, the Eurocode 3 "Design rules for steel structures" consists of 20 parts and Eurocode 4 "Composite construction" of 3 relevant parts. The core forms the so-called basic standard EN 1993-1 "for bases and above

1. INTRODUCTION

ground construction ", which consists again of 11 parts 1-1 to 1-11, to which another part, 1-12 for high strength steel grades (S500 to S700) was added.

In the design standards containing the general rules, two parts are related to fatigue. These are part 1-10: material toughness and through-thickness properties (material quality selection) (EN 1993-1-10:2005), and part 1-9: fatigue (EN 1993-1-9:2005).

Furthermore, the following parts of Eurocode 3 (for definitions of abbreviations see Table 1.1) contain sections on fatigue design of structures which may have to be designed against fatigue:

- Part 1: Steel structures, general rules and rules for buildings EN 1993-1-1)
- Part 2: Steel bridges (EN 1993-2)
- Part 3: Towers and masts and chimneys (EN 1993-3)
- Part 4: Silos and tanks (EN 1993-4)
- Part 6: Crane supporting structures (EN 1993-6).

The same organization holds true for Eurocode 4, for steel and concrete composite structures. Historically, during the revision of ENV-versions into prEN (and thereafter into final versions of EN), the CEN TC250 committee agreed to carry out a reorganization of the rules, including the rules related to fatigue, into the generic and associated Eurocodes. In terms of fatigue in steel and steel and concrete composite structures, it has been agreed that all rules for fatigue were to be compiled and summarized in a new Part 1-9 (in the old versions ENV, they were still in the individual standards) (see Table 1.1). In this new generic part EN 1993-1-9 "fatigue", the rules applicable for the fatigue design of all structures with steel members are regrouped. This part regroups essentially the various so-called "chapter 9" of the former versions of the ENV 1993-1 to ENV 1993-7 parts, see also Sedlacek *et al* (2000). This reorganisation avoids repetition and, in particular, reduces the risk of contradictions between different Eurocode parts. However, not all elements of the fatigue verification are integrated in the new part 1-9. The action effects which are independent from the fatigue resistance are regulated in the EN 1991 parts. Moreover, for some structures, e.g. bridges, towers, masts, chimneys, etc., specific fatigue features remain in the appropriate application parts of EN 1993.

The recommendations in EN 1991-2 on fatigue load models and in EN 1993-1-9 allow for a simplified fatigue verification using fatigue strength (Wöhler, 1860) curves. In addition, a detailed computation with application of the damage accumulation is also possible, allowing for the evaluation of the residual life, for example. In practice, the simplified verification is more user-friendly and more efficient for a daily use.

The limit state of fatigue is characterized by crack propagation followed by a final failure of the structural member. To verify this limit state, a verification of the failure, to avoid brittle fracture of the structural members, is required. The brittle failure is influenced by the material toughness, temperature and thickness. The topic of brittle fracture of steel and the proper choice of material to avoid it, is covered by Eurocode 3, Part 1-10 (EN 1993-1-10) and is presented thoroughly in chapter 6.

1.3.3 Eurocode 9

For aluminium structures, the design codes are only a few in comparison to the steel ones. Eurocode 9 addresses the design of new structures made out of wrought aluminium alloys and gives limited guidance for cast alloys. Eurocode 9 is separated into 5 parts:

- EN 1999-1-1: general structural rules
- EN 1999-1-2: structural fire design
- EN 1999-1-3 : structures susceptible to fatigue
- EN 1999-1-4 : cold-formed structural sheeting
- and EN 1999-1-5: shell structures.

The only part of interest in this book is EN 1999-1-3.

Table 1.1 – Overview and changes in the transition from ENV to EN versions of the various Eurocode 3 and Eurocode 4 parts

ENV-Version	EN-Version	Content
ENV1993-1-1: 1992	EN 1993-1-1: 2005*	General rules and rules for buildings
ENV 1993-1-2: 1995	EN 1993-1-2: 2005*	General rules - Structural fire design
ENV 1993-1-3: 1996	EN 1993-1-3: 2006*	General rules - Supplementary rules for cold-formed members and sheeting

1. INTRODUCTION

Table 1.1 – Overview and changes in the transition from ENV to EN versions of the various Eurocode 3 and Eurocode 4 parts (continuation)

ENV-Version	EN-Version	Content
ENV 1993-1-4: 1996	EN 1993-1-4: 2006	General rules - Supplementary rules for stainless steels
ENV 1993-1-5: 1997	EN 1993-1-5: 2006*	General rules - Plated structural elements
ENV 1993-1-1: 1992	EN 1993-1-6: 2007*	Strength and stability of shell structures
ENV 1993-1-1: 1992	EN 1993-1-7: 2007*	Strength and stability of planar plated structures subject to out of plane loading
ENV 1993-1-1: 1992	EN 1993-1-8: 2005*	Design of joints
ENV 1993-1-1:1992, Chap.9 ENV 1993-2: 1997, Chap.9 ENV 1993-3-1: 1997 ENV 1993-3-2: 1997 ENV 1993-6: 1999	EN 1993-1-9: 2005*	Fatigue
ENV 1993-1-1: 1992, Appendix C ENV 1993-2: 1997, Appendix C	EN 1993-1-10: 2005*	Material toughness and through-thickness properties
ENV1993-2: 1997, Appendix C	EN 1993-1-11: 2006*	Design of structures with tension components
-	EN 1993-1-12: 2007*	General - High strength steels
ENV 1993-2: 1997	EN 1993-2: 2006*	Steel bridges
ENV 1993-3-1: 1997 (prEN 1993-7-1:2003)	EN 1993-3-1: 2006*	Towers, masts and chimneys – Towers and masts
ENV 1993-3-2:1997 (prEN 1993-7-1: 2003)	EN 1993-3-2: 2006	Towers, masts and chimneys – Chimneys

Table 1.1 – Overview and changes in the transition from ENV to
EN versions of the various Eurocode 3 and Eurocode 4 parts (continuation)

ENV-Version	EN-Version	Content
ENV 1993-4-1: 1999	EN 1993-4-1: 2007*	Silos
ENV 1993-4-2: 1999	EN 1993-4-2: 2007*	Tanks
-	EN 1993-4-3: 2007*	Pipelines
ENV 1993-5: 1998	EN 1993-5: 2007*	Piling
ENV 1993-6: 1999	EN 1993-6: 2007*	Crane supporting structures
ENV 1994-1-1: 1992	EN 1994-1-1: 2004	General rules and rules for buildings
ENV 1994-1-2: 1994	EN 1994-1-2: 2005*	Structural fire design
ENV 1994-2: 1997	EN 1994-2: 2005*	General rules and rules for bridges

*corrigenda has been issued for this part.

1.3.4 Execution (EN 1090-2)

The Euronorm EN 1090 fixes the requirements for the execution of steel and aluminium structures, in particular, structures designed according to any of the EN 1993 generic parts and associated Eurocodes, members in steel and concrete composite structures designed according to any of the EN 1994 parts and aluminium structures designed according to EN 1999 parts. EN 1090 is divided in three parts, namely:

- EN 1090-1: Execution of steel structures and aluminium structures – Part 1: general delivery conditions.
- EN 1090-2: Execution of steel structures and aluminium structures – Part 2: Technical requirements for the national execution of steel structures.
- EN 1090-3: Execution of steel structures and aluminium structures – Part 3: Technical requirements for aluminium structures.

The implementation and use of EN 1090 rules is closely linked with the implementation of the structural Eurocodes. The withdrawal of national standards codes in CEN member countries and their replacement by the Eurocodes is now completed in most countries.

1. INTRODUCTION

EN 1090 specifies requirements independently from the type, shape and loading of the structure (e.g. buildings, bridges, plated or latticed elements). It includes structures subjected to fatigue or seismic actions. It specifies the requirements related to four different execution classes, namely EXC1, EXC2, EXC3 and EXC4 (from the less to the more demanding). It is important to note that this classification can apply to the whole structure, to part(s) of it or to specific joints only. Thus, the execution of a building or of any structure would not be, apart from a few exceptions, specified "of execution class 4" as a whole. The particularly severe requirements of this class apply only to certain members, even to only a few essential joints (Gourmelon, 2007).

In order not to leave the design engineer and its client without any clue to answer this question and to avoid the classic, but uneconomical reflex of choosing the most demanding class, guidance for the choice of execution class was elaborated. The principles of the choice are based on three criteria:

- *Consequence classes.* EN 1990: 2002 gives in its Annex B guidelines for the choice of consequence class for the purpose of reliability differentiation. The classification criterion is the importance of the structure or the member under consideration, in terms of its failure consequences. Consequence classes for structural members are divided in three levels, see Table 1.2. The three reliability classes RC1, RC2, and RC3 with their corresponding reliability indexes as given in EN 1990, Annex A1, may be associated with the three consequence classes (Simões da Silva *et al*, 2010).
- *Service categories,* arising from the actions to which the structure and its parts are likely to be exposed to during erection and use (dynamic loads, fatigue, seismic risk, …) and the stress levels in the structural members in relation to their resistance. There are two different possible service categories, see Table 1.3,
- *Production categories,* arising from the complexity of the execution of the structure and its structural members (e.g. complex connections, high strength steels, heavy plates or particular techniques). There are two different possible production categories, see Table 1.4.

1.3 CODES OF PRACTICE

Table 1.2 – Definition of consequence classes (adapted from EN 1990, Table B1)

Cons. Class	Description	Examples of buildings and civil engineering works
CC1	**Low** consequence for loss of human life, *and* economic, social or environmental consequences **small or negligible**	Agricultural buildings where people do not normally enter (e.g. storage buildings, silos less than 100 t capacity, greenhouses)
CC2	**Medium** consequence for loss of human life, economic, social or environmental consequences **considerable**	Residential and office buildings, public buildings where consequences of failure are medium (e.g. office buildings)
CC3	**High** consequence for loss of human life, *or* economic, social or environmental consequences **very great**	Grandstands, public buildings where consequences of failure are high (e.g. concert halls, discretely supported silos more than 1000 t capacity)

The execution classes are then chosen according to the consequence classes, service and production categories determined for the considered members. They can be chosen on the basis of the indications of Table 1.5. Note that in the absence of specification in the contract, execution class 2 applies by default (Gourmelon, 2007).

Table 1.3 – Suggested criteria for service categories (from EN 1090-2, Table B.1)

Cat.	Criteria
SC1	Structures/components designed for quasi static actions only (e.g. buildings) Structures and components with their connections designed for seismic actions in regions with low seismic activity, of low class of ductility (EN 1998-1) Structures/components designed for fatigue actions from cranes (class S_0)*
SC2	Structures/components designed for fatigue actions according to EN 1993 (e.g. road and railway bridges, cranes (classes S_1 to S_9)*) Structures susceptible to vibrations induced by wind, crowd or rotating machinery Structures/components with their connections designed for seismic actions in regions with medium or high seismic activity, of medium or high classes of ductility (EN 1998-1)
* For classification of fatigue actions from cranes, see EN 1991-3 and EN 13001-1

1. INTRODUCTION

Table 1.4 – Suggested criteria for production categories (from EN 1090-2, Table B.2)

Cat.	Criteria
PC1	Non welded components manufactured from any steel grade products
	Welded components manufactured from steel grade products below S355
PC2	Welded components manufactured from steel grade products from S355 and above
	Components essential for structural integrity that are assembled by welding on construction site
	Components with hot forming manufacturing or receiving thermic treatment during manufacturing
	Components of Circular Hollow Sections (CHS) lattice girders requiring end profile cuts

Table 1.5 – Recommendation for the determination of the execution classes (from EN 1090-2, Table B.3)

Consequence classes		CC1		CC2		CC3	
Service categories		SC1	SC2	SC1	SC2	SC1	SC2
Production categories	PC1	EXC1	EXC2	EXC2	EXC3	EXC3*	EXC3*
	PC2	EXC2	EXC2	EXC2	EXC3	EXC3*	EXC4
* EXC4 should be applied to special structures or structures with extreme consequences of a structural failure as required by national provisions							

With emphasis on fatigue behaviour, the execution of welding is of particular importance. Any fault in workmanship may potentially reduce the fatigue strength of a detail. Good workmanship, on the contrary, will result in an increase in the fatigue strength, often above the characteristic S-N curves given in the codes. Indeed, these curves correspond to lower bound test results obtained from average fabrication quality details. Even though good workmanship cannot be quantified in the Eurocodes and used in fatigue verifications, S-N curves referring for most details to failure from undetectable flaws, it can be considered as a welcomed supplementary safety margin.

The good workmanship criteria, however, on which the weld quality specifications in codes and standards are based, are sometimes not directly related to the effect and importance of the feature specified on fatigue strength (or any other strength criteria). Faults in workmanship proven to be detrimental to fatigue include the following (from most to less detrimental, however depending upon original fatigue strength of detail and fault level):

- unauthorised attachments,
- weld lack of fusion/penetration, particularly in transverse butt welds,
- poor fit-up, assembly tolerances, eccentricity and misalignment,
- notches, sharp edges,
- distortion,
- corrosion pitting,
- weld spatter,
- accidental arc strikes.

Speaking again about normative execution requirement, the concern of the design engineer is to choose the class of imperfections tolerated with regards to the reference code EN ISO 5817 (ISO 5817, 2006). In the Eurocode framework and EN 1090-2, the engineer will have, as for the other execution questions, to define the required execution class only. In EN 1090-2, the following requirements are fixed:

- Execution class 1 (EXC1): Quality level D
- Execution class 2 (EXC2): Quality level C
- Execution class 3 (EXC3): Quality level B
- Execution class 4 (EXC4): Quality level B with additional requirements to account for fatigue effects.

For structures or parts of structures which have to be designed against fatigue, quality level D is excluded, quality level C may be used for specific details and quality B is the usual choice. It should be noted that a fatigue detail category 90 seems to be compatible with most imperfections of quality level B according to ISO 5817, with however the exception of the following, where more stringent requirements should be set (Hobbacher et al, 2010):

- continuous undercut
- single pore, pore net, clustered porosity
- slag inclusions, metallic inclusions
- linear misalignment of circumferential welds
- angular misalignment (which is not in ISO 5817 at this time)
- multiple imperfections in longitudinal direction of weld.

A project for revising ISO 5817, in particular with respect to fatigue criteria, is under discussion. The allowable imperfections requirements go

1. INTRODUCTION

along with requirements on the company quality system, welding coordination, etc. A summary of the main welding requirements is given in Table 1.6. It can be seen that there is no requirements for EXC1, which once again shows it is not adequate for structures under fatigue loadings.

Table 1.6 – Main weld requirements, extracts from EN 1090-2

	EXC1	EXC2	EXC3	EXC4
Qualif. of welding procedures; welding coordination	Not required.	Required, see EN 1090-2, § 7.4	Required, see EN 1090-2, § 7.4	Required, see EN 1090-2, § 7.4
Temporary attachments	Not req.	Not req.	Use to be specified Cutting and chipping not permitted	
Tack welds	Not req.	Qualified welding procedure		
Butt welds	Not req.	Run on/off pieces if specified	Run on/off pieces For single side welds, permanent backing continuous	
Execution of welding	Not req.	Not req.	Removal of spatter	
Acceptance criteria	EN ISO 5817 Quality level D if specified	EN ISO 5817 Quality level C generally	EN ISO 5817 Quality level B	EN ISO 5817 Quality level B+

The additional requirements for quality B+ are given in EN 1090-2 Table 17. In summary, this table gives additional or more severe limits for imperfections such as undercut (not permitted in B+), internal pores, linear misalignment, etc. It also gives supplementary requirements for bridge decks.

Outside of the welding requirements, it is very important to meet every special requirement in order not to impair fatigue strength. Thus, it should be emphasised that all connections provided for temporary structural members or for fabrication purposes shall also meet the requirements of EN 1090. Also, regarding member identification, a suitable system shall be put into place in order to be able to follow each piece; note that the suitable marking method is function of the material. Furthermore, the marking methods shall be applied in a way not producing damage and only on areas where it does not affect the fatigue life.

Finally, note that in addition to the rules found in EN 1090, some additional information and requirements regarding execution can also be found directly in the detail category tables of EN 1993-1-9, as well as in the other EN 1993 and EN 1994 relevant parts for the different types of structures. For example, in Table 8.5 of EN 1993-1-9, detail 1 to 3, cruciform and tee joints, the following requirement is given: *the misalignment of the load-carrying plates should not exceed 15 % of the thickness of the intermediate plate.*

1.3.5 Other execution standards

With emphasis on fatigue behaviour, the standards related to welding, bolting and erection are the most important ones. The complete list of standards in these areas, about 200 "normative References", is given in EN 1090. The largest group are the standards for products, for which there are around one hundred. Then, there are about thirty standards related to welding, about fifteen dealing with destructive or non-destructive testing applicable to the welds, as well as about twenty standards in relation with corrosion protection (Gourmelon, 2007). To mention only a few, of relevance to the subject presented in this book:

- EN ISO 3834: 2005 (in 6 parts), Quality requirements for fusion welding of metallic materials. This standard defines requirements in the field of welding so that contracting parties or regulators do not have to do it themselves. A reference to a particular part of EN ISO 3834 should be sufficient to demonstrate the capabilities of the manufacturer to control welding activities for the type of work being done. The different parts of the standards deal with the following items: contract and design review, subcontracting, welding personnel, inspection, testing and examination personnel, equipment, storage of parent materials, calibration, and identification/traceability.
- EN ISO 5817: 2003 (corrected version 2005), Welding – fusion-welded joints in steel, nickel, titanium and their alloys – quality levels for imperfections. This standard defines the dimensions of typical imperfections, which might be expected in normal fabrication. It may be used within a quality system for the production of factory-welded joints. It provides three sets of dimensional values (quality levels B, C, and D) from which a selection can be made for a particular application. This standard is directly applicable to visual testing of

welds and does not include details of recommended methods of detection or sizing by non destructive means. It does not cover metallurgical aspects, such as grain size or hardness.
- EN ISO 9013: 2002, Thermal cutting – Classification of thermal cuts – Geometrical product specification and quality tolerances. This standard applies to materials and thickness ranges suitable for oxyfuel flame cutting, plasma cutting and laser cutting.
- EN 12062: 1997, Non-destructive testing of welds – General rules for metallic materials. The purpose of this standard is quality control. It gives guidance for the choice and evaluation of the results of non-destructive testing methods based on quality requirements, material, weld thickness, welding process and extent of testing. This standard also specifies general rules and standards to be applied to the different types of testing (visual inspection, dye-penetrant flaw detection, eddy-current tests, magnetic-particle flaw detection, radiographic testing, and ultrasonic testing), for either the methodology or the acceptance level for metallic materials.
- ISO 15607: 2003: Specification and qualification of welding procedures for metallic materials - General rules. This standard gives the general rules and requirements concerning the qualification of welding procedures, which are further developed in EN 15611, EN 15612, EN 15613, and EN 15614.

1.4 DESCRIPTION OF THE STRUCTURES USED IN THE WORKED EXAMPLES

1.4.1 Introduction

In order to fulfil the objectives of the design manuals of this ECCS collection, three different structures were chosen to be used for the detailed design examples presented in this book. Before being used for fatigue calculations, they are introduced and briefly described in the following paragraphs. The first structure is a steel and concrete composite bridge which is also used in the ECCS design manual about EN 1993 part 1-5 (plate buckling) (Beg *et al*, 2010). The second structure considered is a chimney and the third one is a crane supporting structure.

1.4 DESCRIPTION OF THE STRUCTURES USED IN THE WORKED EXAMPLES

1.4.2 Steel and concrete composite road bridge (worked example 1)

1.4.2.1 Longitudinal elevation and transverse cross section

This worked example is adapted from the COMBRI research project (COMBRI, 2007) (COMBRI+, 2008). The bridge is a symmetrical composite box-girder structure with five spans, 90 m + 3 × 120 m + 90 m (i.e. a total length between abutments equal to 540 m, see Figure 1.12). It is assumed to be located in Yvelines, near Paris, France. For simplification reasons, the horizontal alignment is assumed straight as well as the road, the top face of the deck is horizontal and the structural steel depth is constant and equal to 4000 mm.

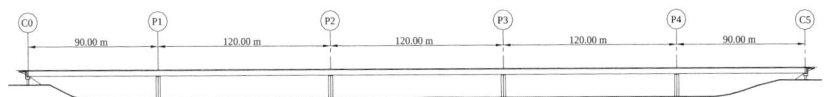

Figure 1.12 – Side view of road bridge with span distribution

A four-lane traffic road crosses the bridge. Each lane is 3.50 m wide and the two outside ones are bordered by a 2.06 m wide safety lane. Normalised safety barriers are located outside the traffic lanes and in the centre of the road (see Figure 1.13). The cross section of the concrete slab and non-structural equipments is symmetrical with respect to the axis of the bridge. The 21.50 m wide slab has been modelled with a constant thickness equal to 0.325 m. The slab span between the main girders is equal to 12.00 m and the slab cantilever is 4.75 m on both sides.

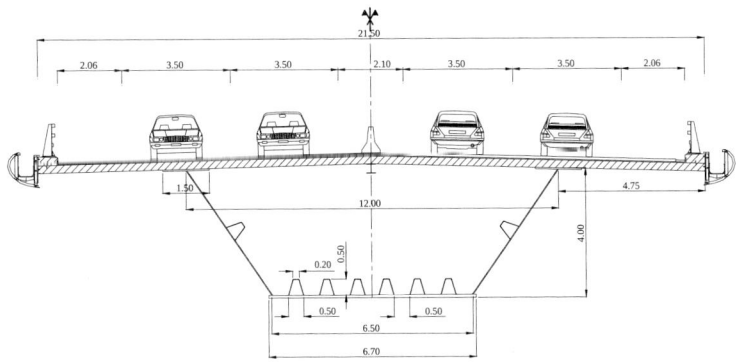

Figure 1.13 – Cross section of road bridge with lane positions

The concrete slab is connected to an open box-section. The centre-to-centre distance between webs in the upper part is equal to 12.00 m, and to

1. INTRODUCTION

6.50 m in the lower part. The upper flanges are 1500 mm wide whereas the lower flange is 6700 mm wide.

1.4.2.2 Materials and structural steel distribution

A steel grade S355 has been used. Its mechanical properties are given in EN 10025-3 and slightly modified by EN 1993-2 (see Table 1.7). Normal concrete of class C35/45 is used for the reinforced concrete slab; the reinforcing steel bars are class B high bond bars with a yield strength of 500 MPa and a modulus of elasticity equal to 210000 MPa (as structural steel).

Table 1.7 – f_y and f_u according to the plate thickness for steel grade S355

t (mm)	f_y (MPa)	f_u (MPa)	Quality
$t < 16$	355	470	K2
$16 \leq t < 30$	345	470	N
$30 \leq t < 40$	345	470	NL
$40 \leq t < 63$	335	470	NL
$63 \leq t < 80$	325	470	NL
$80 \leq t < 100$	315	470	NL
$100 \leq t$	295	450	NL

The structural steel distribution results from a design according to Eurocodes 1, 3 and 4, see Figure 1.15.

The thickness variations of the upper and lower flanges are found towards the inside of the girder. Due to the concrete slab width, an additional plate welded to each upper flange (1400 mm wide and welded below the main one) needs to be added in the regions of intermediate supports.

Cross frames stiffen the box-section on abutments and on intermediate supports, as well as every 4.0 m in the spans, see Figure 1.14. The bottom flange longitudinal trapezoidal stiffeners are continuous with a plate thickness equal to 15 mm, see Figure 1.16. The web longitudinal stiffeners are discontinuous; these have the same thickness throughout and are located at mid-depth to provide sufficient cross section shear resistance. An additional longitudinal steel rolled I-girder (located right in the middle of the bridge cross section) spans between the transverse frames and is directly connected to the concrete slab. It helps for the slab concreting phases and resists with the composite cross section (as an additional section for the upper steel flanges).

1.4 DESCRIPTION OF THE STRUCTURES USED IN THE WORKED EXAMPLES

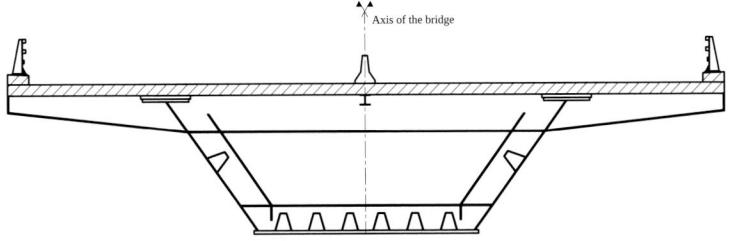

Figure 1.14 – Cross frame on supports

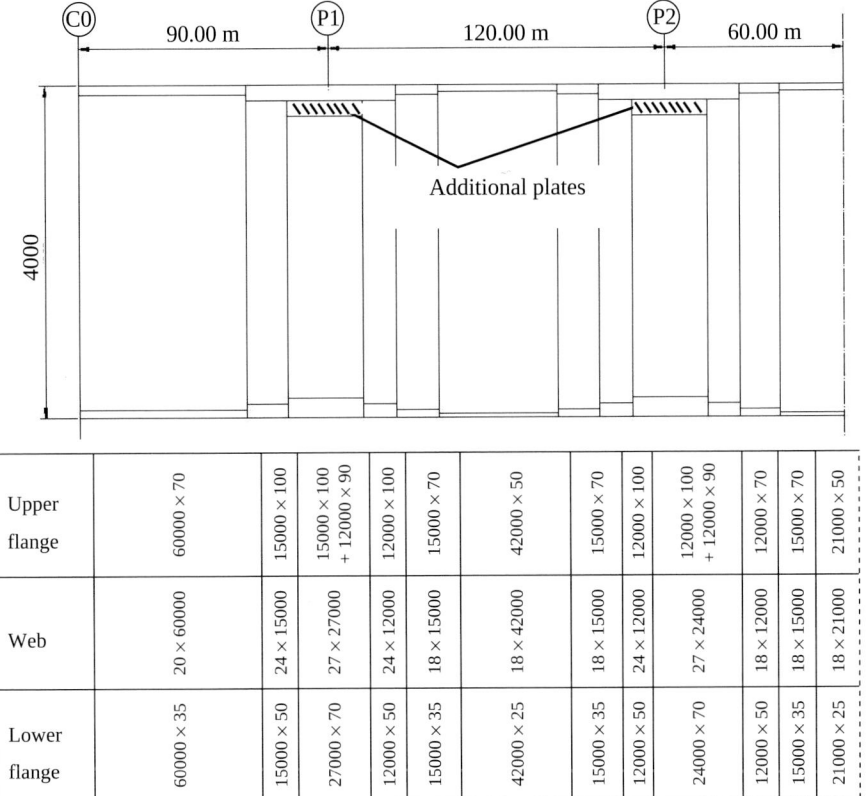

Figure 1.15 – Structural steel distribution (half length of the road bridge)

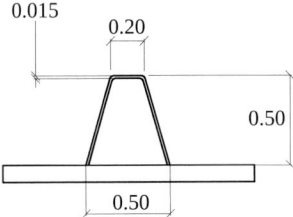

Figure 1.16 – Detailed view of longitudinal stiffener

1. Introduction

1.4.2.3 The construction stages

The assumptions pertaining to the construction stages should be taken into account when calculating the internal moments and forces distribution in the bridge deck (EN1994-2, 5.4.2.4) as they have a high influence on the steel/concrete modular ratios. For the bridge example, the following construction stages have been adopted:

- Launching of the structural steel structure
- On-site pouring of the concrete slab segments by casting them in a selected order (first the in-span segments, and second the segments around internal supports):

    The total length of the bridge (540 m) has been broken down into 45 identical 12-m-long concreting segments. The start of pouring the first slab segment is the time of origin ($t = 0$). The time taken to pour each slab segment is assessed at 3 working days. The first day is devoted to the concreting, the second day to its hardening and the third to move the mobile formwork. The slab is thus completed within 135 days.

- The installation of non-structural equipments is assumed to be completed within 35 days, so that the deck is fully constructed at the date $t = 135 + 35 = 170$ days.

1.4.3 Chimney (worked example 2)

1.4.3.1 Introduction

The following worked example is based on a real chimney verification made by Kammel (2003) and the ECCS Technical Committee 6. The original example was published in Stahlbaukalendar (2006) and has since been adapted by the authors since the original verification was carried out to determine the cause of observed cracks. Indeed, the existing chimney had fatigue problems due to vortex shedding induced vibrations. The example presented includes a tuned mass damper in order to solve this problem.

Vortex shedding often occurs in cantilevered steel chimneys that are subjected to dynamic wind loads. At a critical wind speed, alternating vortices detach from the cylindrical shell over a specific correlation length causing a vibration of the structure transverse to the wind direction, see section 3.1.5 for further explanations. The cyclic loading is transferred to all

1.4 DESCRIPTION OF THE STRUCTURES USED IN THE WORKED EXAMPLES

structural members and connections. In this type of structure, the following structural details are usually relevant for fatigue verification:

- Bolted flange connection between two sections,
- Welded stiffeners at the bottom,
- Anchor bolts at the bottom
- Inspection manholes and/or inlet tubes details.

The chimney dealt with in this example has a height of 55 m and an outside constant diameter of 1.63 m, as shown in Figure 1.17. It is a double-walled chimney with an outer tube and inner thermal insulating layer. The chimney shaft is composed of 5 separate parts, bolted together using socket joints, see Figure 1.20. At the bottom, a reinforced ring is used and the chimney is held down using 28 anchor bolts, see Figure 1.18, Figure 1.19 and Figure 1.21. Furthermore, an inspection manhole is present in the bottom zone, see Figure 1.19. All dimensions and other information are given in the next paragraphs.

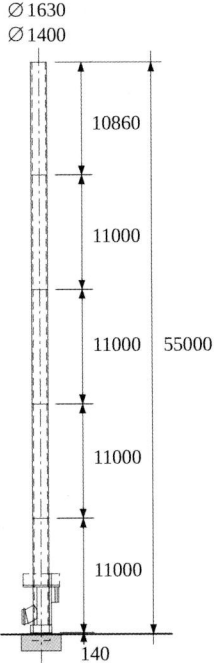

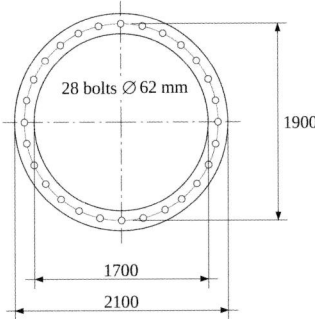

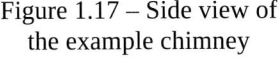

Figure 1.17 – Side view of the example chimney

Figure 1.18 – Anchor bolts at +0.350 m (plan view)

1. Introduction

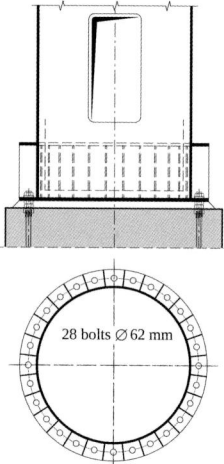

Figure 1.19 – Drawing of bottom part of chimney with manhole position, section and top view, ground plate with anchor bolts at +0.350 m

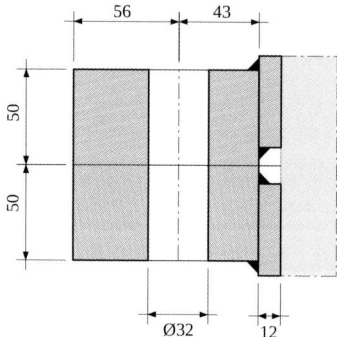

Figure 1.20 – Relevant bolted flange connection between two sections at +11.490 m (remark: this flange design corresponds to standard practice and does not represent optimum design; an improved design is possible)

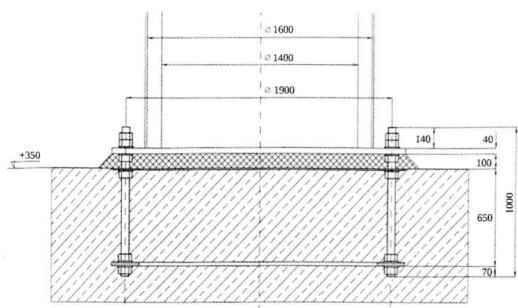

Figure 1.21 – Ground plate with anchor bolts at +0.350 m (section view)

1.4 DESCRIPTION OF THE STRUCTURES USED IN THE WORKED EXAMPLES

1.4.3.2 General characteristics of the chimney

Height	: $h = 55.0$ m
Outer diameter	: $b = 1630$ mm
Slenderness ratio	: $\lambda = h/b = 33.7$
Shell thickness from bottom up to +11.490m	: $s = 12$ mm
Shell thickness at top	: $s' = 6$ mm
Steel yield stress	: $f_y = 190$ N/mm²

(which corresponds to steel S235 operating at a max. temperature $T = 100$ °C)

In the choice of material quality, since tensile stresses occur perpendicular to the ring surface at the joint between ring and shell (socket joint, see Figure 1.20), attention has to be paid to avoid lamellar tearing and the designer must follow the rules given in EN 1993-1-10.

The equivalent total mass per unit length is taken as $m_e = 340$ kg/m (EN 1991-1-4, Section F.4: for cantilevered structures with a varying mass distribution m_e may be approximated by the average value of m over the upper third of the structure).

For damping characteristics, the logarithm decrement is taken as (EN 1991-1-4, Section F.5)

$$\delta = \delta_s + \delta_a + \delta_d = 0.03$$

This value includes damping from a tuned mass damper. For the structural part only, for a welded chimney without external thermal insulation, the value given in EN 1991-1-4 is lower, $\delta_s = 0.012$. A tuned mass damper is a type of dynamic vibration absorber which must be specifically analysed, tested and tuned on the structure and periodically inspected (EN 1993-3-2, annex B). Examples of chimneys with vibration problems and their resolution are periodically published, for example (Kawecki et al, 2007).

1.4.3.3 Dimensions of socket joint located at +11.490 m (see Figure 1.20)

Bolt diameter (M30; 10.9)	: $D_{M30} = 30$ mm
Bolt cross section	: $A_{s,30} = 561$ mm²

1. Introduction

Bolt resistance	: f_{ub} = 1000 N/mm²
Bolt preload	: $F_{p,Cd}$ = 350 kN
Total number of bolts	: n = 42
Distance of bolts and shell	: a = 43 mm
Distance between bolts	: $e = \frac{\pi}{n} \cdot (b + 2 \cdot a) = 128.4$ mm
Washer dimensions	: t_{was} = 5 mm
	d_a = 56 mm
	d_i = 31 mm
Socket flange cross section	: t_f = 50 mm
	w = 99 mm
	$b' = w - a = 56$ mm

Section of the chimney (without socket):

$$A_K = \frac{\pi}{4} \cdot \left(b^2 - (b - 2 \cdot s)^2\right) = 61000 \, \text{mm}^2$$

Elastic section modulus $\quad : W_y = \frac{\pi}{32} \cdot \frac{b^4 - (b - s \cdot 2)^4}{b} = 24493 \cdot 10^3 \, \text{mm}^3$

1.4.3.4 Dimensions of ground plate joint with welded stiffeners located at the bottom, at +0.350 m:

Anchor bolt diameter (M60; 8.8)	: D_{M60} = 60 mm
Anchor bolt cross section	: $A_{s,60}$ = 2362 mm²
Anchor bolt resistance	: f_{ub} = 800 N/mm²
Total number of anchor bolts	: n = 28 (same as number of stiffeners)
Distance of bolts and shell	: a = 135 mm
Radius of anchor bolt circle	: r_s = 1900 / 2 = 950 mm
Distance between bolts	: $e = \frac{\pi}{n} \cdot (b + 2 \cdot a) = 213.2$ mm
Longitudinal stiffener	: 646 × 223 × 10 mm
Upper ring stiffener	: 250 × 25 mm

1.4 DESCRIPTION OF THE STRUCTURES USED IN THE WORKED EXAMPLES

Double-sided fillet welding between stiffener and ground plate: $a_w = 6$ mm

If flux-cored welding is used, the effective weld throat is larger than a_w, and this value can be used in the verifications. We will assume here that $a_{w,eff} = a_w + 1$ mm.

	$L_w = 220$ mm
Ground plate dimensions	: $d_{bp} = 2100$ mm
	$t_{bp} = 40$ mm

1.4.3.5 Dimensions of manhole located betweeen +1.000 m and +2.200 m:

Height	: $h_{mh} = 1200$ mm
Width	: $w_{mh} = 600$ mm
Corner radius	: $r_{mh} = 300$ mm
Opening sides reinforcement plates	: $b_{rp} = 90$ mm
	$t_{rp} = 10$ mm
	$h_{rp} = 1400$ mm

Section of the chimney (with manhole):

$$A_{K,mh} = \frac{\pi}{4} \cdot \left(b^2 - (b-2\cdot s)^2\right) - w_{mh} \cdot s + 2 \cdot t_{rp} \cdot b_{rp} = 55600 \text{ mm}^2$$

Assumed elastic section modulus at manhole level:

$$W_{y,mh} = 20000 \cdot 10^3 \text{ mm}^3$$

1.4.4 Crane supporting structures (worked example 3)

1.4.4.1 Introduction

Figure 1.22 presents the general geometry of the single crane supporting structure example. This example is adapted from one presented in TGC 11 (2006). The crane supporting structure is composed of a continuous runway beam – HEA 280 in S 355 steel - supported by surge girders, with spans between supports $l = 6$ m. It is assumed that the end-spans are shorter, thus the relevant span is an inner span. The crane span is $s = 14.30$ m and the nominal hoist load is $Q_{nom} = 100$ kN.

1. Introduction

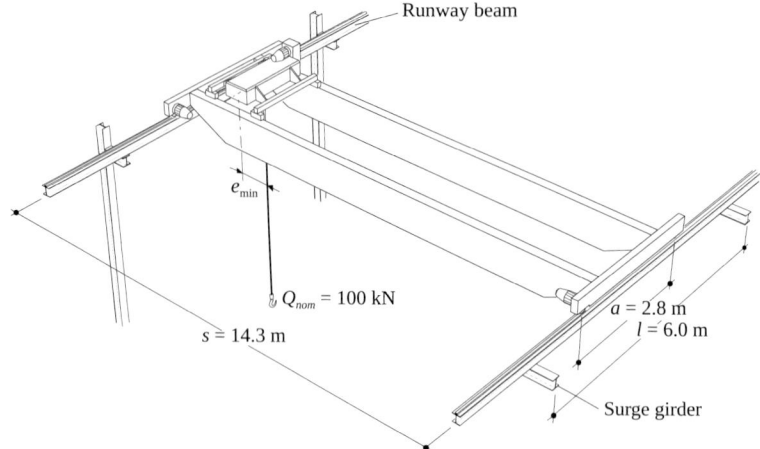

Figure 1.22 – Crane supporting structure

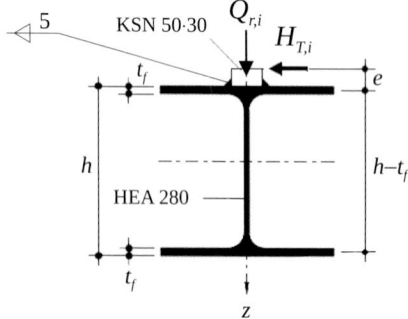

Figure 1.23 – Cross section of the runway beam

The single crane supporting structure is classified according to section 2.12, of EN 1991-3 (the classification table can also be found in EN 13001-1) as a function of the total number of lifting cycles and the load spectrum as follows:

- Class of load spectrum: Q4
- Class of total number of cycles: U4

1.4.4.2 Actions to be considered

The following characteristic values of actions are considered as provided by the crane supplier:

Nominal hoist load : $Q_{nom} = 100$ kN
Maximum load per wheel of loaded crane : $Q_{r,max} = 73.4$ kN

1.4 Description of the Structures Used in the Worked Examples

Minimum load per wheel of unloaded crane	: $Q_{r,min}$ = 18.75 kN
Horizontal transverse load per wheel	: $H_{T,i}$ = 9.4 kN
Self weight runway beam (HEA 280 + KSN 50×30, see Figure 1.23)	: g_k = 88.2 kg/m·10 m/s² = 0.882 kN/m
Neutral axis position of runway beam measured from bottom fiber (weared rail)	: z_g = 149 mm
Inertia of runway beam with weared rail (for static considerations, according to TGC11 (2006))	: I_y = 155.8 · 10^6 mm⁴
Inertia of runway beam's effective upper flange and weared rail (for fatigue considerations, 12,5% of the rail height)	: I_y = 189.8 · 10^3 mm⁴

Chapter 2

APPLICATION RANGE AND LIMITATIONS

2.1 INTRODUCTION

The fatigue strength curves and detail categories given in Eurocode 3 part 1-9 are mainly based on fatigue tests carried out on bolted and welded carbon steels with nominal yield stress ranging from 235 to 400 N/mm^2, i.e. mainly S235 and S355 steels. Under the condition of non-corrosive environmental conditions, numerous studies have shown that the rules in EN 1993-1-9 could be applied to other steel grades and steel types, including stainless steel alloys. In other words, the influence of steel grade and steel type can be neglected compared to the influence of detailing and weld imperfections. Also, the fatigue strength curves given in part 1-9 apply only to structures operating under normal atmospheric conditions and with sufficient corrosion protection and regular maintenance.

The application field embraces also structural members from EN 1993-1-11 that is pre-stressing bars, ropes and cables.

Part 1-9 is not applicable to:

- Oligo-cyclic or low-cycle fatigue, that is when a few cycles cause fatigue fracture (e.g. earthquake) or, more generally, when nominal direct (normal) stress ranges exceed 1.5 f_y or nominal shear stress ranges exceed 1.5 $f_y/\sqrt{3}$. An additional condition, expressed in the IIW recommendations, refers also cases where hot-spot stress σ_{hs} exceeds 2 f_y. This can be the case for pressure vessels, tanks or silos.
- Structures subjected to temperatures exceeding 150 °C (e.g. pressure vessels, pipework).
- Structures in corrosive media (gases, liquids) other than normal atmospheric conditions.

2. Application Range and Limitations

- Materials not behaving in a ductile manner, not conforming to the toughness requirements of EN 1993-1-10, see chapter 6.
- Structures in seawater environment (e.g. offshore structures).
- Structures subjected to single impact.
- Concrete reinforcement, steel rebars.

2.2 MATERIALS

Part 1-9 covers the structural steel grades and connecting devices listed in EN 1993-1-1, sections 3.2 and 3.3, with extension to higher structural steel grades given in EN 1993-1-12, namely:

- Structural steel grades S235 to S700, according to EN 10 025, EN 10 149, EN 10 210 and EN 10 219.
- Austenitic and Duplex stainless structural steels according to EN 10 088. Note that although there are differences in mechanical behaviour between structural steel and structural stainless steel alloys, it has been shown that the fatigue curves and rules for ferritic steels can be applied to welded stainless steel alloys (excluding environmental considerations).
- Structural steels with improved atmospheric corrosion resistance, according to EN 10 025-5.

Part 1-9 may be used for other structural steels, provided that adequate and sufficient data exist to justify the application. It only applies to materials which conform to the toughness requirements of EN 1993-1-10, see chapter 6. The influence of corrosion on fatigue strength is developed in the next section.

2.3 CORROSION

Severe corrosion acts like sharp notches, considerably reducing the lifetime of the structure under fatigue loading. Normal steel grades must therefore have adequate corrosion protection such as:

- Paint systems according to ISO 12944 (1998);
- Hot-dip galvanizing (with care to avoid steel embrittlment (Feldmann et al, 2008) (Pargeter, 2003);

- Cathodic protection;
- Self-protecting layers such as the one developing on weathering and stainless steels.

Weathering steel grades can be left unprotected in mild corrosive environments such as structures exposed to rain washing and sun drying, free of salt, where details do not trap debris, do not stay wet for long periods of time, and are regularly maintained (acid rain is not considered an especially severe condition). The protective oxide layer that develops is however rougher than the surface of a normal carbon steel and thus reduces the fatigue strength for the higher fatigue classes. In other words, for plain member details in classes 160, 140 and 125 (details 1 to 5 from Table 8.1, EN 1993-1-9), the next lower category must be used if the detail is made with weathering steel, that is categories 140, 125 and 112, respectively, instead of the original ones. In all other cases, the details can be classified into standard detail categories since slight corrosion notches have less influence than the geometric imperfections or the welding produced notches.

Stainless steels do not have the problem of weathering steels and all detail categories are the same as for corrosion protected carbon steels.

Nevertheless, it should be emphasised that welding in addition to excessive corrosion notches reduce severely the fatigue strength of all types of steel. Special attention regarding corrosion protection should be given under the following circumstances:

- steel structures in marine environments (250-500 m from the sea), or subjected to salt-laden fogs,
- where run-off from de-icing salt reaches the structure and is not washed off by rain,
- where there are highly corrosive chemicals or industrial fumes in the atmosphere.

2.4 TEMPERATURE

The effects of temperature on the fatigue strength of a detail should be checked. Generally speaking, it has been shown for structural steels and aluminium alloys that there is no significant change in fatigue crack growth rates with low temperatures, down to service temperatures of $-50\ °C$, unless

2. Application Range and Limitations

brittle fracture becomes the governing crack propagation mode (Schijve, 2001). Thus, for structural steels, a proper choice of material to avoid brittle fracture is sufficient in most cases and the effects of low temperatures do not need to be further considered in fatigue design and verification. In particular cases of exposure to low temperatures such as structures in arctic regions and cold storage cells for example, specific studies should be carried out and high quality steels should be used.

In this book, temperatures above 150 °C are not considered as they fall outside of the scope of EN 1993-1-9. The onset of the influence of high temperatures on fatigue strength strongly depends upon the material and other environmental conditions. For example, the fatigue strength of a stainless NiCrMo steel is not affected until 400 °C (Schijve, 2001), as for structural aluminium alloys it is already affected at 70 °C (IIW, 2009). Since a reduction in the fatigue strength for structural steels can occur at temperatures exceeding 100 °C, a conservative design approach is recommended. As a general rule, it can be said that the reduction of the fatigue strength is proportional to the ratio between the elastic modules at service and room temperature. If the elastic modulus at the service temperature is not known, the reduced fatigue strength, $\Delta\sigma_{C,temp}$, can be computed using the following formula (IIW, 2009):

$$\frac{\Delta\sigma_{C,temp}}{\Delta\sigma_C} = 1.045 - 290 \cdot 10^{-6} \cdot T - 1.3 \cdot 10^{-6} \cdot T^2 \qquad (2.1)$$

where
T temperature in Celsius.

The use of such a reduced fatigue strength rule due to high temperature effects cannot be made systematically. In some domains such as pressure vessels and pipelines, some specific rules exist. EN 1993-3-2 contains the following rule for the influence of high temperature on fatigue behaviour of towers and combined effects of temperature and applied stresses: for chimneys made of heat resistant alloy steels which are used at temperatures above 400 °C, the addition of the temperature induced damage with the fatigue damage should be duly accounted for. The temperature limit beyond which the material stability deteriorates should be determined, and fundamental understanding of the material behaviour is important. High temperature fatigue problems require experimental research while knowledge of material science is indispensable for planning research (Schijve, 2001). The authors thus recommend using a design by testing approach as explained in sub-chapter 4.3.

2.5 LOADING RATE

The loading frequency has, up to approximately 100 Hz, no influence on the fatigue behaviour of steel structures. However, under the combined effects of cyclic loading and corrosion, or cyclic loading and high temperatures, the loading frequency has a significant influence on crack propagation. But since these combinations are excluded from the scope of EN 1993-1-9, the standard does not include any information on loading frequencies. In cases of combined effects, the engineer is advised to use a design by testing approach, see sub-chapter 4.3.

2.6 LIMITING STRESS RANGES

According to EN 1993-1-9, the stress ranges (nominal, corrected nominal, or structural stress at the hot spot) under the frequent combination of action effects $\psi_1 \cdot Q_k$ (see EN 1990), are limited to a maximal value of $1.5 f_y$ under direct stresses (according to the terminology used in EN 1993-1-9) and to $1.5 f_y / \sqrt{3}$ under shear stresses. The maximal possible stress range $\Delta\sigma = 2 \cdot f_y$ is limited to the value $1.5 f_y$. In design, ultimate static strength limit states will usually govern, so that this criteria, is of secondary importance, except for hybrid girders.

The limitation of the stress range to the value $1.5 \cdot f_y$ refers to the region of oligo-cyclic fatigue (up to max 50 000 load cycles, see section 1.1.3), where the use of the fatigue strength equation would lead to values highly exceeding the yield stress, or the ultimate strength. From the stress range limitation one can deduce also a limiting number of cycles value through the relationship (2.2):

$$N \geq 2 \cdot 10^6 \left(\frac{\Delta\sigma_C}{1.5 f_y} \right)^3 \qquad (2.2)$$

2. APPLICATION RANGE AND LIMITATIONS

For the detail category 160 and a steel grade S235, this relationship leads to $N \geq 187000$ cycles. This value is superior to the 50000 cycles mentioned before; it represents an upper bound value for limiting number of cycles between short life and long life. This limitation between short life and long life is of importance in the domains of pressure vessels, silos and tanks; these structures are subjected to a small number of cycles, but of very high magnitude. In EN 1993-4-1 (silos) for example, for silos classified in consequence classes 2 and 3, e.g. large silos (see section 1.3.4) it is required that parts of the structure subjected to severe bending should be checked against fatigue and cyclic plasticity limit states using the procedures given in EN 1993-1-6 and EN 1993-1-7 as appropriate. Silos of small capacities (consequence class 1) are excluded from any fatigue or cyclic plasticity verifications.

A limit on the stress range is also necessary for hybrid girders. Recall that hybrid girders are structural members where, usually, two different grades of steels are combined. Typically, one can find such girders in the shape of I-plate girders in bridges. Because of the higher stresses they are subjected to, the design calls for flanges made of a high strength steel grade, while for the web, because of plate stability limitations, a lower steel grade is used. Under static loading, as shown in Figure 2.1, the part of the web next to the tension flange will yield before the latter. At the static ultimate limit state, this early yielding of the web has no influence, as long as the web and the weld between the flange and the web have sufficient ductility. One can just mention that the resulting deflections, compared with a girder made entirely with the high strength steel grade, will be somewhat larger at the service limit state.

Under fatigue loading, the yielding and shakedown of the self-equilibrated stresses occurring during the first cycles in a hybrid girder will be similar to those occurring in a "single steel grade" girder. Since fatigue strength of welded members is independent of steel grade, the fatigue strength of a hybrid girder is identical to that of a girder made out of a single material. However, there is still the 1.5 times yield stress limit.

Since there are different yield stress values, it has been shown that the limit on the stress range in the web, $\Delta\sigma_{Web}$, can be set in function of the yield stress of the flange material, $f_{y,Flange}$ (and not the one from the web). EN 1993-1-12 specifies that for flanges grades greater than S460 up to S700 fulfilling condition $f_{y,Flange} \leq \pi f_{y,Web}$ according to EN 1993-1-5 4.3 (6) the maximal stress range value is thus limited to:

2.6 LIMITING STRESS RANGES

$$\Delta\sigma_{Web} \leq 1.5 f_{y,Flange} \tag{2.3}$$

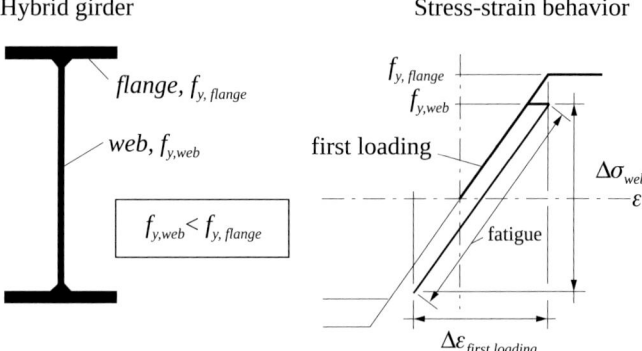

Figure 2.1 – Hybrid girder stress-strain fatigue cycles

Chapter 3

DETERMINATION OF STRESSES AND STRESS RANGES

3.1 FATIGUE LOADS

3.1.1 Introduction

For structures subject to fluctuating stresses, the fatigue loads are represented by fatigue load models and/or groups of loads and their number of occurrences, as given in the relevant parts of Eurocode 1 - Actions on Structures. Fatigue load models may differ from the load models of the Ultimate Limit State (ULS) and Serviceability Limit State (SLS). Depending upon the type of structure, a fatigue load can either be given as a moving load (like trucks on a bridge) or as a load range acting at a fixed location (wind on a mast).

In Eurocode 1, fatigue loads are given as:

- standardized load groups Q_i and their corresponding frequencies of occurrence or n_i,
- a maximum or damage equivalent constant load Q_E and its frequency of occurrence or n_{max},
- a damage equivalent constant load $Q_{E,2}$ related to $2 \cdot 10^6$ load cycles.

Dynamic effects are to be accounted for in the definition of the fatigue load models. Different cases are presented in the following paragraphs according to the type of structure.

For typical fatigue-stressed structures, Table 3.1 contains the relevant Eurocode standards in which the fatigue loads are defined. For other structures, see indications in section 3.1.8.

3. DETERMINATION OF STRESSES AND STRESS RANGES

Table 3.1 – Overview of Eurocode standards to assess the fatigue strength of different structures

Structure	Relevant fatigue actions from	Standard	Related standard load or load range
Road bridges	Road vehicles	EN 1993-2	EN 1991-2
Railway bridges	Trains	EN 1993-2	EN 1991-2
Crane supporting structures	Cranes	EN 1993-6	EN 1991-3
Masts and towers of chimneys	Wind	EN 1993-3-1	EN 1991-1-4
Silos and tanks	Loads from filling and emptying	EN 1993-4-1 EN 1993-4-2	EN 1991-4

3.1.2 Road bridges

For road bridges, the fatigue loads are defined in EN1991-2. Overall, five different fatigue load models denoted FLM1 to FLM5 are given. These models correspond, in principle, to various uses, in so far as it was decided, from inception, that the Eurocode should give (Calgaro *et al*, 2009):

– one or more rather pessimistic loads models to quickly identify in which parts of the structure a problem of fatigue could appear,
– one or more models to perform usual simple verifications,
– one or more models to perform accurate verifications (based on damage accumulation calculation).

The above, as well as the format of fatigue verification used in the associated Eurocodes will decide which model is more appropriate. The associated Eurocodes are: EN1993-2 for a steel bridge and EN1994-2 for a composite steel and concrete bridge.

The models FLM1 and FLM2 are used to verify that the bridge lifetime is infinite regarding the fatigue phenomena. It means that no fatigue crack propagation could occur in any structural detail of the bridge. Such verification requires the definition of a constant amplitude fatigue limit (CAFL), which is not always the case (for instance the S-N curves for shear have no CAFL). In the model FLM2, only one lorry travelling on the slow lane of the bridge is considered. It can be used if the effect of a second (or more) lorry on the bridge deck can be neglected and then FLM2 is more precise than FLM1.

The models FLM3, FLM4 and FLM5 are used to verify that the bridge has a correct lifetime, coherent with the project assumptions, regarding the fatigue phenomena. The verification should be based on the S-N curves defined in the different Eurocodes. Each model is aimed at representing the whole traffic, as a single vehicle (model FLM3) in a very simple way, as a set of equivalent lorries (model FLM4), or as a traffic registration (model FLM5) in a very precise (but complex) way. If the effect of a second (or more) lorry on the bridge deck can be neglected, the model FLM4 is more precise than the model FLM3.

3.1.2.1 Fatigue load model 1 (FLM1)

This model derives from the principal characteristic load model for road bridges, called LM1 and defined in EN1991-2. LM1 is set up with a uniform design load (UDL) and tandem system (TS). For the traffic lane i, FLM1 is the superposition of 0.7 Q_{ik} (for characteristic TS) and 0.3 q_{ik} (for the characteristic uniform design load). FLM1 is thus close to LM1 frequent values and as it is defined, it is very conservative (Calgaro *et al*, 2009).

In practice, this model is not called for by the generic parts or associated Eurocodes. This is due to the fact that the frequent SLS combination of actions, used for other verifications than fatigue and already calculated, is very similar and defined with the frequent LM1, (i.e. $0.75Q_{ik} + 0.4q_{ik}$). For instance, the criterion for oligo-cyclic fatigue defined in EN 1993-1-9, 8(1), is based on this frequent SLS combination. In this case, there is no need for specifying a number of cycles.

3.1.2.2 Fatigue load model 2 (FLM2)

This model is a set of frequently idealised lorries which are defined in EN 1991-2 (Table 4.6) and reproduced below, Table 3.2. Each lorry has a frequent axle load value, Q_{ik}, and should cross the bridge alone in the appropriate slow traffic lane. The aim is to determine a maximum stress range (for the set of lorries) that should be compared to the CAFL. With such a verification, there is no need for specifying a number of cycles. As this approach requires a very reliable calibration (not yet performed), this model 2 is not called for by the different associated Eurocodes.

3. Determination of Stresses and Stress Ranges

Table 3.2 – Set of frequent lorries for road bridges, fatigue load model 2 (source EN 1991-2, Table 4.6)

1	2	3	4
LORRY SILHOUETTE	Axle spacing (m)	Frequent axle loads (kN)	Wheel type (see Table 4.8, EN 1991-2)
	4.50	90 190	A B
	4.20 1.30	80 140 140	A B B
	3.20 5.20 1.30 1.30	90 180 120 120 120	A B C C C
	3.40 6.00 1.80	90 190 140 140	A B B B
	4.80 3.60 4.40 1.30	90 180 120 110 110	A B C C C

3.1.2.3 Fatigue load model 3 (FLM3)

This simplified fatigue load model consists of a 4 axles single vehicle with a weight of $Q_E = 120$ kN per axle (see Figure 3.1). Its use is associated with the concepts of the equivalent stress range at 2 millions cycles and damage equivalent factors (see section 3.2), so that the fatigue load model 3 is of high practical importance for engineers.

The model 3 crosses the bridge in the mid-line of the slow traffic lane defined in the project. A second 4 axles vehicle, with a reduced load of 36 kN per axle, can follow the first one with a minimum distance equal to 40 m. This can govern the fatigue design of a structural detail located on an intermediate bridge support, each adjacent span being loaded by one of the two lorries.

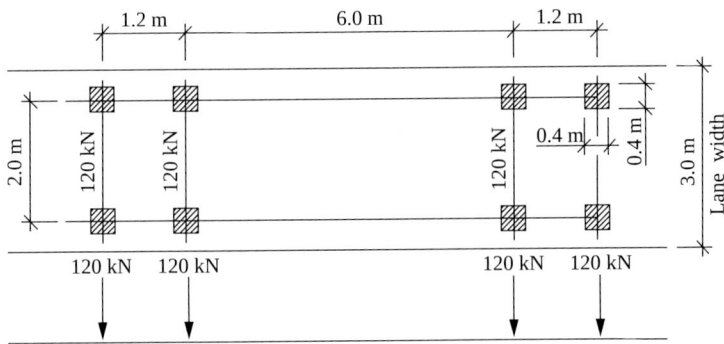

Figure 3.1 – Fatigue load model 3 for a road bridges according to EN 1991-2

Since verification can be made with respect to finite fatigue life, there is a need for specifying a number of cycles, which is expressed as a traffic category on the bridge. A traffic category should be defined by at least (EN 1991-2):

- the number of slow lanes,
- the number N_{obs} of heavy vehicles (maximum gross vehicle weight more than 100 kN), observed or estimated, per year and per slow lane (i.e. a traffic lane used predominantly by lorries).

Indicative values are given in EN 1991-2 and reproduced in Table 3.3, but the national annexes may define traffic categories and numbers of heavy vehicles.

Table 3.3 – Indicative numbers of heavy vehicles expected per year and per slow lane (source EN 1991-2)

	Traffic categories	N_{obs} per year and per slow lane
1	Roads and motorways with 2 or more lanes per direction with high flow rates of lorries	$2.0 \cdot 10^6$
2	Roads and motorways with medium flow rates of lorries	$0.5 \cdot 10^6$
3	Main roads with low flow rates of lorries	$0.125 \cdot 10^6$
4	Local roads with low flow rates of lorries	$0.05 \cdot 10^6$

3. Determination of Stresses and Stress Ranges

On each fast lane (i.e. a traffic lane used predominantly by cars), additionally, 10 % of N_{obs} may be taken into account.

It should be noted that there is no general relation between traffic categories for fatigue verifications and the ultimate strength loading classes and associated adjustment factors. The establishment of the FLM3 was performed using data from measurements of real traffic on the motorway Paris-Lyon at Auxerre, France (Sedlacek *et al*, 1984). Within this fatigue load model are already included effects of the flowing traffic, such as the pavement quality and dynamic responses of the bridges.

3.1.2.4 Fatigue load model 4 (FLM4)

This model is a set of five "equivalent" lorries. Each lorry is assumed to cross the bridge alone and represents a certain percentage of the heavy traffic according to the road type (long distance, medium distance, or local traffic), see Table 3.4. This FLM4 model also needs the definition of N_{obs} which is the total number of lorries crossing the bridge per year. It can be taken from the indicative percentage values given in EN 1991-2 or should be defined in the bridge specifications according to the road type or traffic measurements. Its use is associated with the verification using the damage accumulation method, see sections 1.1.4 and 5.4.4.

Table 3.4 – Set of equivalent lorries for road bridges, fatigue load model 4 (source EN 1991-2, Table 4.7)

VEHICLE TYPE (corresponding wheel type, see Table 3.2)			TRAFFIC TYPE AND LORRY PERCENTAGE		
1	2	3	4	5	6
LORRY SILHOUETTE	Axle spacing (m)	Frequent axle loads (kN)	Long distance	Medium distance	Short distance
	4.50	70 130	20	40	80
	4.20 1.30	70 120 120	5	10	5

Table 3.4 – Set of equivalent lorries for road bridges, fatigue load model 4 (source EN 1991-2, Table 4.7) (continuation)

VEHICLE TYPE			TRAFFIC TYPE AND LORRY PERCENTAGE		
(corresponding wheel type, see Table 3.2)					
1	2	3	4	5	6
LORRY SILHOUETTE	Axle spacing (m)	Frequent axle loads (kN)	Long distance	Medium distance	Short distance
	3.20 5.20 1.30 1.30	70 150 90 90 90	50	30	5
	3.40 6.00 1.80	70 140 90 90	15	15	5
	4.80 3.60 4.40 1.30	70 130 90 80 80	10	5	5

3.1.2.5 Fatigue load model 5 (FLM5)

This model is the most general one and consists of registered traffic data. Its use is associated with statistical tools, first to identify and count the stress ranges (using the reservoir or rainflow methods) and secondly to extrapolate the bridge fatigue life from short registered period(s) and from a set of assumptions regarding the future traffic evolution.

3.1.3 Railway bridges

Similarly to the road bridges, the fatigue loads for railway bridges are defined in EN 1991-2. However, in the case of railway bridges for the fatigue verification using the damage equivalence concept, a special fatigue load model was not developed. Instead, the characteristic fatigue loads are obtained from the static load model 71 (or SW/0 in case of continuous beams

3. DETERMINATION OF STRESSES AND STRESS RANGES

or SW/2 for heavy loads), as shown in Table 3.5, but excluding the ultimate strength design adjustment factor. The fatigue load model has to be placed on each of the tracks; thus the internal forces obtained correspond to the maximum effects of all tracks loaded.

The influence of the train speed, the structure rigidity, the track quality as well as other different influences are considered by the dynamic factor Φ_2. This simplified procedure for the dynamic effects is not valid for high speed trains (> 200 km/h). In this case, a specific dynamic analysis is required for paying attention to the fatigue stress ranges induced by vibrations or resonance phenomena.

The fatigue assessment should be carried out on the basis of the traffic mixes: "standard traffic", "traffic with 250 kN-axles" or "light traffic mix", depending on whether the structure carries standard traffic mix, predominantly heavy freight traffic or light traffic (EN 1991-2). The derivation of the damage equivalent factors is based on exactly defined traffic compositions: suburban traffic and goods traffic, which are indicated informatively in the EN 1991-2 Annex D with 12 different train types and their daily frequencies and weights. Numbers of trains or cycles are given but the damage equivalent factors do not refer to those as it was found not to be the most relevant parameter. Instead, the traffic volume is used. The indicative value of standard traffic is 25 million of tons per year and per track.

Table 3.5 – Load model 71 and SW/0

Type	Geometry and load diagram
Load model 71	Q_{vk} = 250 kN 250 kN 250 kN 250 kN; q_{vk} = 80 kN/m; q_{vk} = 80 kN/m; unlimited 0.8 m 1.6 m 1.6 m 1.6 m 0.8 m unlimited
SW/0	q_{vk}; a; c; a a = 15 m ; c = 5.3 m ; q_{vk} = 133 kN/m
SW/2	q_{vk}; a; c; a a = 25 m ; c = 7.0 m ; q_{vk} = 150 kN/m

3.1.4 Crane supporting structures

For crane supporting structures, the fatigue loads are given in EN 1991-3. The crane variable loads are primarily due to the variation of the lifted load and the movement of the crane along the runway beam. For the fatigue verification, the static load model should be used as there is no specific fatigue load model defined. Figure 3.2 gives the load arrangements to obtain the relevant vertical actions to the runway beam. Cranes loading classification for fatigue can be made according to the recommendations given in EN 1991-3, Annex B - Table B1, which makes the link between the service classes (S_i, see Table 3.6) and the hoisting classes (HC_i)

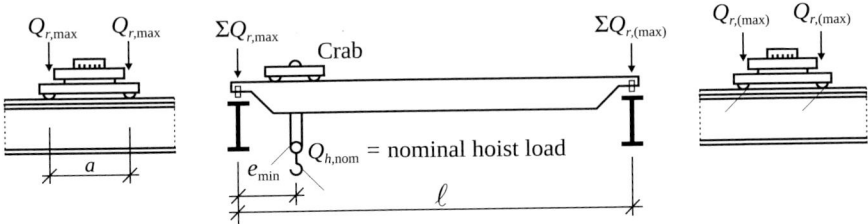

a) Load arrangement of the loaded crane to obtain the maximum loading on the runway beam.

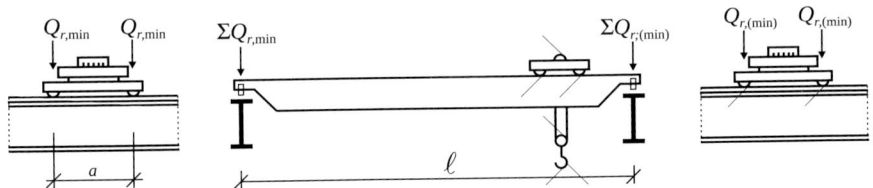

b) Load arrangement of the loaded crane to obtain the minimum loading on the runway beam.

Figure 3.2 – Load arrangements to obtain the relevant vertical actions on the runway beams (from EN 1991-3)

For normal service condition of the crane the fatigue loads may be expressed in terms of damage equivalent fatigue load Q_E (that may be taken as constant for all crane positions to determine fatigue load effects). However, in the case of fatigue verifications, specific dynamic impact factor values φ_{fat} are given, which differ from the values for the bearing capacity limit state. The damage equivalent fatigue load Q_E is given in equation (3.1).

$$Q_E = \varphi_{fat} \cdot Q_{max,i} \qquad (3.1)$$

3. Determination of Stresses and Stress Ranges

where

$Q_{max,i}$ maximum value of the characteristic vertical wheel load i,

φ_{fat} damage equivalent dynamic impact factor (EN 1991-3, clause 2.12.1(7))

Detailed information on the new European rules for crane supporting structures can be found in Kuhlmann et al (2003). The number of stress cycles can be higher than the number of crane working cycles (EN 1993-6, figure 9.1), as shown in Figure 3.3 for the local stresses due to wheel passages (see sub-section 5.4.7.3). In these cases, according to 2.12.1(4) of EN 1991-3, the damage equivalent fatigue load should be used in conjunction with this higher number as the total number of working cycles (Table 2.11 of EN 1991-3).

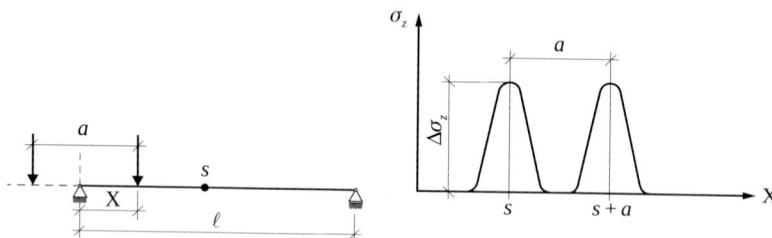

Figure 3.3 – Two stress cycles rising from one crane working cycle (source EN 1993-6)

Table 3.6 – Classification of the fatigue actions from cranes into service classes according to EN 13001-1

Class of load spectrum		Q_0	Q_1	Q_2	Q_3	Q_4	Q_5
		$kQ \leq 0.0313$	$0.0313 < kQ \leq 0.0625$	$0.0625 < kQ \leq 0.125$	$0.125 < kQ \leq 0.25$	$0.25 < kQ \leq 0.5$	$0.5 < kQ \leq 1.0$
class of total number of cycles							
U_0	$C \leq 1.6 \cong 10^4$	S_0	S_0	S_0	S_0	S_0	S_0
U_1	$1.6 \times 10^4 < C \leq 3.15 \times 10^4$	S_0	S_0	S_0	S_0	S_0	S_1
U_2	$3.15 \times 10^4 < C \leq 6.30 \times 10^4$	S_0	S_0	S_0	S_0	S_1	S_2
U_3	$6.30 \times 10^4 < C \leq 1.25 \times 10^5$	S_0	S_0	S_0	S_1	S_2	S_3
U_4	$1.25 \times 10^5 < C \leq 2.5 \times 10^5$	S_0	S_0	S_1	S_2	S_3	S_4
U_5	$2.5 \times 10^5 < C \leq 5.00 \times 10^5$	S_0	S_1	S_2	S_3	S_4	S_5

Table 3.6 – Classification of the fatigue actions from cranes into service classes according to EN 13001-1 (continuation)

Class of load spectrum		Q_0	Q_1	Q_2	Q_3	Q_4	Q_5
		$k_Q \leq$ 0.0313	0.0313 $< k_Q \leq$ 0.0625	0.0625 $< k_Q \leq$ 0.125	0.125 $< k_Q \leq$ 0.25	0.25 $< k_Q \leq$ 0.5	0.5 $< k_Q \leq$ 1.0
class of total number of cycles							
U_6	$5.00 \times 10^5 < C \leq 1.00 \times 10^6$	S_1	S_2	S_3	S_4	S_5	S_6
U_7	$1.00 \times 10^6 < C \leq 2.00 \times 10^6$	S_2	S_3	S_4	S_5	S_6	S_7
U_8	$2.00 \times 10^6 < C \leq 4.00 \times 10^6$	S_3	S_4	S_5	S_6	S_7	S_8
U_9	$4.00 \times 10^6 < C \leq 8.00 \times 10^6$	S_4	S_5	S_6	S_7	S_8	S_9

where:
kQ   is a load spectrum factor for all tasks of the crane;
C   is the total number of working cycles during the design life of the crane.
Note: The classes S_i are classified by the stress effect history parameter *s* in EN 13001-1 which is defined as:
$s = \upsilon \, k$ where:
k   is the stress spectrum factor;
υ is the number of stress cycles *C* related to 2.0×10^6 stress cycles.
The classification is based on a total servisse life of 25 years.

3.1.5 Masts, towers, and chimneys

For slender buildings like masts, towers and chimneys, evidence of fatigue due to wind is becoming increasingly important. It is the subject of recent research activities, see for example Schaumann and Seidel (2001), Schmidt and Jakubowski (2001) and Verwiebe (2003). Niemann and Peil (2003) makes an overview of fatigue due to wind. The two main wind vibration mechanisms are, according to Niemann and Peil (2003):

– self excited (intrinsic) vibrations, which are in the wind direction (i.e. along-wind vibrations) and
– vortex-induced vibrations, which occur perpendicular to the wind direction (e.i. cross-wind vibrations).

The along-wind vibrations occur because the force exerted by the wind varies with the wind speed and its associated turbulence. Their intensities depend upon the natural frequency of the structure and structure-wind interaction. This is a broad-band process which leads to a spectrum of different stress ranges with a relatively small total number of cycles (in comparison to vortex-shedding).

In case of two or more structures, wake interferences may occur. The most common vibrations a chimney downstream of other chimneys can experience are buffeting and galloping. Buffeting is defined as the unsteady loading of a structure by velocity fluctuations in the incoming flow and not self-induced. Buffeting vibration is the along-wind vibration produced by turbulence (Simiu and Scanlan, 1986). Galloping is caused by variation in drag and lift forces in the wake of a chimney that can lead to cross-wind oscillations in the downstream chimney. Single flexible structures with non-circular cross sections or sections on which ice has formed are also prone to galloping as a self-induced cross-wind vibration mode.

For assessing fatigue loading from along-wind vibrations, see EN 1993-3-1, section 9.2.1. The number of load cycles for gust response, N_g, is given in EN 1991-1-4, annex B.3. The relationship links a wind effect value ΔS (pressure range, force range, or displacement range) to the number of times it is reached or exceeded during a period of 50 years, see equation (3.2). Furthermore, the relationship is normalized as a percentage of the characteristic wind force value S_k.

$$\frac{\Delta S}{S_k} = 0.7 \cdot \left(\log N_g\right)^2 - 17.4 \cdot \log N_g + 100 \qquad (3.2)$$

where
  S_k wind force, or its effect, due to a 50 years return period wind action,
  ΔS wind force range occurring or exceeded N_g times during service life (50 years).

As for vortex-shedding, this phenomenon occurs when vortices are shed alternately from opposite sides of the structure. This gives rise to a fluctuating load perpendicular to the wind direction (i.e. cross-wind vibrations). This is a narrow-band process which leads to a very large number of cycles of the same range. In EN 1993-3-2, a note is given which can be taken as general rule; fatigue from cross-wind vortex vibrations normally governs chimney design, thus the fatigue verification related to inline forces need normally not to be undertaken. Apart from the wind effects, the dynamic system performance (e.g. damping) is also of great importance. The vortex-induced amplitudes may be

3.1 FATIGUE LOADS

reduced by means of aerodynamic devices (only under special conditions, e.g. Scruton numbers larger than 8) or damping devices supplied to the structure (tuned mass dampers, etc.).

For the case of the vortex-induced lateral vibrations, a procedure for the determination of the stress range spectrum is given in EN 1991-1-4, Appendix E. For light steel structures, the lateral vibrations usually occur with the same amplitude (by opposition to concrete structures), thus the load spectrum is narrow-banded and the load spectrum factor can be taken as a constant (load spectrum factor $k_Q = 1.0$). The value of the stress range, $\Delta\sigma$, is computed from the maximum system deflection amplitude $y_{F,max}$. For the determination of this maximum system deflection, two different procedures are indicated in EN 1991-1-4; they use the dynamic characteristics such as the Strouhal number and the Reynolds number.

The relevant number of load cycles caused by vortex shedding, N_v, is a function of the frequency of occurrence of the critical wind velocity for vortex-induced vibration. According to EN 1991-1-4, the number of vibration cycles is to be determined in accordance with the relationship (3.3), which is based upon a Weibull type distribution for the frequency of occurrence of the critical wind velocities per year and under the assumption of a 50 year service life.

The number of load cycles N_v caused by vortex-induced vibration is given by (EN 1991-1-4, section E.1.5.2.6, expression E.10):

$$N_v = 2 \cdot T \cdot n_y \cdot \varepsilon_0 \cdot \left(\frac{v_{crit}}{v_0}\right)^2 \cdot \exp\left(-\left(\frac{v_{crit}}{v_0}\right)^2\right) \quad (3.3)$$

where
- n_y natural frequency of cross-wind mode [Hz]. Approximations for n_y are given in EN 1991-1-4, Annex F,
- v_{crit} critical wind velocity in [m/s] given in EN 1991-1-4, annex E.1.3.1,
- v_0 $\sqrt{2}$ times the modal value of the Weibull probability distribution assumed for the wind velocity [m/s],
- T service life in seconds, which is equal to $3.2 \cdot 10^7$ multiplied by the expected service life in years,
- ε_0 bandwidth factor describing the range of wind velocities with vortex-induced vibrations (in the range 0.1 to 0.3. It may be conservatively taken as $\varepsilon_0 = 0.3$).

3. Determination of Stresses and Stress Ranges

In the case of guyed masts, the fatigue performance of guys should be verified using the procedures given in EN 1993-1-11 for cables, see sections 3.1.7 and 4.2.10. The example of a steel chimney subjected to wind loading is now presented.

Example 3.1: Application to a chimney (worked example 2), computation of wind loads for fatigue from vortex shedding.

In this application, only the case of vortex shedding induced vibrations and fatigue is checked (cross-wind vibrations). In certain cases, one should also check gust winds fatigue (along-wind vibrations). Wind loads on the chimney described in section 1.4.3 are now computed according to EN 1991-1-4.

From sub-section 1.4.3.2, the chimney has the following dimensions:

$h = 55$ m

$b = 1630$ mm

$\lambda = \dfrac{h}{b} = 33.7$

an equivalent total mass per unit length of $m_e = 340$ kg/m

a structural damping logarithmic decrement of $\delta_s = 0.012$

and a logarithmic decrement given by $\delta = 0.03$

Terrain roughness

The chimney is located in a suburban terrain, in a flat area with regular cover of buildings; thus it is classified as a terrain category III (EN 1991-1-4, annex A).

The roughness factor, $c_r(z)$ (EN 1991-1-4, section 4.3.2), is given by:

$$c_r(z) = \begin{cases} k_r \ln\left(\dfrac{z}{z_0}\right) & z_{min} \leq z \leq z_{max} \\ c_r(z_{min}) & z \leq z_{min} \end{cases}$$

where

z_0 roughness length

k_r terrain factor depending on the roughness lenght z_0

The roughness length z_0 is given in Table 4.1 of EN 1991-1-4, section 4.3.2, as $z_0 = 0.3$ m with $z_{min} = 5$ m and $z_{0,II} = 0.05$ m

The terrain factor, k_r, is given by:

$$k_r = 0.19 \cdot \left(\frac{z_0}{z_{0,II}}\right)^{0.07} = 0.19 \cdot \left(\frac{0.3}{0.05}\right)^{0.07} = 0.266$$

The chimney satisfies the criteria for checking vortex shedding at section where vortex shedding occurs (i.e. $h/b > 6$ and $v_{crit,1} \leq 1.25 v_m$, EN 1991-1-4, section E.1.2).

<u>Dynamic characteristics of the chimney (EN 1991-1-4, Annex F)</u>

Fundamental flexural frequency (1st mode) (EN 1991-1-4, section F.2)

$$n_1 = \frac{\varepsilon_1 \cdot b}{h_{eff}^2} \sqrt{\frac{W_s}{W_t}} = \frac{1000 \cdot 1.63}{50.5^2} \sqrt{0.84} = 0.586 \, \text{Hz}$$

where

h_{eff} is the effective height, taken as the total height: $h_{eff} = 50.5$ m

W_s/W_t structural to total weight ratio: $W_s/W_t = 0.84$

ε_1 is equal to 1000 for steel chimneys

Critical wind velocity for bending vibration mode 1 (EN 1991-1-4, section E.1.3.1):

$$v_{crit,1} = b \cdot \frac{n_1}{St} = 1.63 \cdot \frac{0.586}{0.180} = 5.30 \, \text{m/s}$$

where St is the Strouhal number for cylindrical cross sections (EN 1991-1-4, section E.1.3.2), $St = 0.180$

Scruton number (EN 1991-1-4, section E.1.3.3)

$$Sc = \frac{2 \cdot \delta_s \cdot m_e}{\rho \cdot b^2} = \frac{2 \cdot 0.012 \cdot 340}{1.25 \cdot 1.63^2} = 2.46$$

where

ρ is the air density (EN 1991-1-4, section E.1.3.3), $\rho = 1.25$ kg/m^3

$m_e = 340$ kg/m (equivalent total mass per unit length)

3. Determination of Stresses and Stress Ranges

Reynolds number (EN 1991-1-4, section E.1.3.4)

$$Re(v_{crit,1}) = \frac{b \cdot v_{crit,1}}{v} = \frac{1.63 \cdot 5.3}{15 \cdot 10^{-6}} = 5.76 \cdot 10^5$$

where v is the cinematic air viscosity (EN 1991-1-4, section E.1.3.4), $v = 15 \cdot 10^{-6} \, m^2/s$

Cross-wind amplitude (EN 1991-1-4, section E.1.5.2)

Calculation of the cross-wind amplitude $y_{F,max}$ using approach 1 (general approach). The effective correlation length for the 1st vibration mode, L_1, (see Figure 3.4), depends on the vibration amplitude $y_{F,max}$ according to EN 1991-1-4, section E.1.5.2.3:

$$\begin{cases} \dfrac{y_{F,max}}{b} < 0.1 & \Rightarrow \dfrac{L_1}{b} = 6 \\[2mm] 0.1 \leq \dfrac{y_{F,max}}{b} < 0.6 & \Rightarrow \dfrac{L_1}{b} = 4.8 + 12 \cdot \dfrac{y_{F,max}}{b} \\[2mm] \dfrac{y_{F,max}}{b} \geq 0.6 & \Rightarrow \dfrac{L_1}{b} = 12 \end{cases}$$

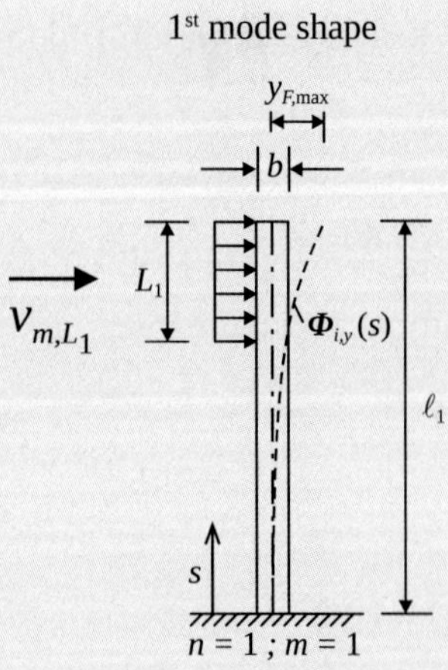

Figure 3.4 – Chimney 1st mode correlation length (EN 1991-1-4)

Note: In reality, the movement is perpendicular to the wind direction.

Adopting an iterative procedure, assuming for the 1st iteration:

3.1 Fatigue Loads

$$\frac{y_{F,max}}{b} < 0.1 \Rightarrow L_1 = 6 \cdot b = 9.78 \text{m}$$

The effective correlation length factor, K_w, can be approximated by (EN 1991-1-4, section E.1.5.2.4):

$$K_w = 3 \cdot \frac{L_1}{b \cdot \lambda} \cdot \left[1 - \frac{L_1}{b \cdot \lambda} + \frac{1}{3} \cdot \left(\frac{L_1}{b \cdot \lambda}\right)^2\right]$$

$$= 3 \cdot \frac{9.78}{1.63 \cdot 33.7} \cdot \left[1 - \frac{9.78}{1.63 \cdot 33.7} + \frac{1}{3} \cdot \left(\frac{9.78}{1.63 \cdot 33.7}\right)^2\right] = 0.444$$

The mode shape factor, K, (EN 1991-1-4, section E.1.5.2.5 and Table E.5, section E.1.5.2.4) is given by:

$$K = 0.130$$

The basic value for the lateral exciting force coefficient, $c_{lat,0}$, (EN 1991-1-4, Section E.1.5.2.2) is given by Figure E.2 as a function of the Reynolds number.

$$c_{lat,0} = 0.200$$

The lateral exciting force coefficient, c_{lat}, (EN 1991-1-4, section E.1.5.2.2), is given in table E.3 as a function of $v_{crit,1}/v_{m,L}$.

Since the mean wind speed in the centre of the effective correlation is given by (EN 1991-1-4, Section 4.3):

$$v_{m,L} = c_r \cdot c_0 \cdot v_b = 1.09998 \cdot 1.0 \cdot 28 = 30.8 \text{ m/s}$$

where

c_r is the roughness factor:

$$c_r(z) = k_r \ln\left(\frac{z}{z_0}\right) = 0.266 \ln\left(\frac{49.55}{0.3}\right) = 1.09998$$

with $z = 49.55$ m (value at the centre of the effective correlation length)

c_0 is the orography factor (equal to 1.0 by default: no hills, slopes, etc)

v_b basic wind velocity (= 28 m/s)

3. Determination of Stresses and Stress Ranges

$$\frac{v_{crit,1}}{v_{m,L}} \leq 0.83 \Rightarrow c_{lat} = c_{lat,0} = 0.200$$

Largest lateral vibration amplitude at $v_{crit,1}$ (EN 1991-1-4, Section E.1.5.2.1):

$$y_{F,max} = \frac{b \cdot K_w \cdot K \cdot c_{lat}}{St^2 \cdot Sc} = \frac{1360 \cdot 0.444 \cdot 0.13 \cdot 0.200}{0.180^2 \cdot 2.46} = 237 \, mm$$

or

$$\frac{y_{F,max}}{b} = \frac{0.237}{1.63} = 0.145 > 0.1$$

Performing a few iterations until convergence gives:

	Iterations			
	1st	2nd	3rd	4th
$y_{F,max}/b$	9.78	10.66	10.87	10.91
L_1	< 0.1	0.145	0.155	0.158
$y_{F,max}$		0.237	0.253	0.258
z				49.55

So that the final value of the effective correlation length is $L_1 = 10.91$ m and the largest lateral vibration amplitude at $v_{crit,1}$:

$$y_{F,max} = \frac{b \cdot K_w \cdot K \cdot c_{lat}}{St^2 \cdot Sc} = 258 \, mm$$

Calculation of the range of exciting forces

The computation of the range of exciting force can be done using a detailed analysis of generalized forces and vibration modes. The bending moment range is given by

$$F_w(s) = m(s) \cdot (2 \cdot \pi \cdot n_{i,y})^2 \cdot \Phi_{i,y}(s) \cdot y_{F,max}$$

where

    $m(s)$ vibrating mass of the structure per unit length (kg/m)
    s location
    $\Phi_{i,y}(s)$ mode shape of the structure normalized to 1 at the point with the maximum displacement
    $n_{i,y}$ natural frequency of the structure

3.1 FATIGUE LOADS

However, one can use simplified methods such as EN 1991-1-4 or Petersen (2000). Both methods give in this case similar results.

Range of resultant exciting force, from Petersen (2000), page 628

$$\Delta P_{lat} = 2 \cdot \left(\frac{\rho}{2} \cdot v_{crit,1}^2 \cdot c_{lat} \cdot b \cdot L_1 \right) = 125.1 \, \text{N}$$

Note: the method from EN 1991-1-4, Annex E.1.4, formula E.6 leads to a similar, however smaller value, $\Delta P_{lat} = 118.9$ N.

Resonance amplification factor, from Petersen (2000), page 628

$$V = \frac{\pi}{\delta} = \frac{\pi}{0.03} = 105$$

Bending moment range at + 11490 mm

$$\Delta M_1 = V \cdot \Delta P_{lat} \cdot \left(h - 11.49 - \frac{L_1}{2} \right) \cdot 1.02 = 508.3 \, \text{kNm}$$

In order to account for 2nd order effects, a simplified constant amplification factor value equal to 1.02 is used. This amplification factor can be estimated using EN 1993-3-2, section 5.2.3. In this example, it could be neglected.

Bending moment range at + 2200 mm (top of manhole)

$$\Delta M_2 = V \cdot \Delta P_{lat} \cdot \left(h - 2.20 - \frac{L_1}{2} \right) \cdot 1.02 = 632.6 \, \text{kNm}$$

Bending moment range at + 1000 mm (bottom of manhole)

$$\Delta M_3 = V \cdot \Delta P_{lat} \cdot \left(h - 1.00 - \frac{L_1}{2} \right) \cdot 1.02 = 648.6 \, \text{kNm}$$

Note: the bending moments at top and bottom of manhole reinforcements, if any (located at + 2300 mm and + 900 mm), are assumed for simplification identical to those at manhole.

Bending moment range at foundation level + 350 mm

$$\Delta M_4 = V \cdot \Delta P_{lat} \cdot \left(h - 0.35 - \frac{L_1}{2} \right) \cdot 1.02 = 657.3 \, \text{kNm}$$

Comment: The same calculation can be made for the second vibration mode. For a very slender chimney ($h/b > 30$), which is this case, the second mode is as important as the first mode. It is usually the governing vibration mode for the fatigue verification of the upper connection due to the larger vibration amplitudes near the top. Furthermore, one must consider a lower damping for the second vibration mode. The example will however concentrate on verifying the lower connections, thus using the 1st mode of vibration only.

The number of cycles resulting from vortex shedding induced vibrations is computed in Example 3.5.

3.1.6 Silos and tanks

In accordance with EN 1991-4, the effects of fatigue are to be considered for silos and tanks if, and only if, the structure under design is subjected to more than one load cycle per day (average value). A load cycle corresponds thereby to a complete filling and emptying of the silo or tank.

In the case of bulk silos, one must consider that they are often equipped with special devices for emptying the silo by vibration of an eccentric motor. Since "bridge" formation is possible in the bulk, the vibration may cause the content of the silo to fall with violence. In each case, one has to examine whether the result of the dynamic effects may lead to a fatigue problem and if so, the silo should be designed accordingly.

3.1.7 Tensile cable structures, tension components

Tension components are the scope of one specific standard: part 1-11 of Eurocode 3. This part is limited to components that are adjustable and replaceable, according to the component list given for fatigue strength, see Table 4.3. These components are generally prefabricated products delivered to site and installed into the structure. However, the rules may also be applied to tension components that are not adjustable or replaceable, e.g. air spun cables of suspension bridges, or for externally post-tensioned bridges. The issue of fatigue is comparatively of greater concern in bridges than in roofs.

Action effects linked to fatigue on tension components are:

3.1 FATIGUE LOADS

- Preload ; needed to avoid detensionning and resulting uncontrolled movement and damages to the component or the structure,
- Wind and rain induced oscillations,
- Live loads,
- Corrosion.

The fatigue strength of tension components is strongly influenced by the possible bending in a tension component, by corrosion actions and their combined effects (corrosion-fatigue), all of which occur essentially near the component ends. Those are used to determine what is called the *exposure class* of the component, see Table 3.7.

Table 3.7 – Determination of the exposure class of tension components

Fatigue action	Corrosion action	
	not exposed externally	exposed externally
no significant fatigue action	class 1	class 2
mainly axial fatigue action	class 3	class 4
axial and lateral fatigue actions (wind & rain)	-	class 5

For exposure classes 3, 4 or 5, fatigue verifications should be carried out using the fatigue actions from EN 1991 and the appropriate category of structural detail, see section 4.2.10.

Regarding fatigue actions, one specific and important case for tension components is vibrations. Aerodynamic forces on a cable may be caused by (see EN 1991-1-4):

a) buffeting (from turbulence in the air flow),
b) vortex shedding (from von Karman vortexes in the wake behind the cable),
c) galloping (self induction),
d) wake galloping (fluid-elastic interaction of neighbouring cables),
e) interaction of wind, rain and cable.

3.1.8 Other structures

EN 1993-4-3 deals with pipelines. In pipelines, cyclic loading can be divided into two classes according to the limit state reached: low cycle fatigue (filling, emptying cycles) or high cycle fatigue (along-wind and cross-wind vibrations). More information can be found in the Eurocodes and in the literature, as for example in Murphy and Langner (1985).

EN 1993-5 deals with the design of steel piling. For structures subjected to cyclic loadings, proper design of connections between the structures and the piles shall be made and adequate corrosion protection provided. Even if the code does not state it, fatigue analyses may be necessary when piles are driven or vibrated into the subsoil.

Apart from the types of structures treated in the associated Eurocodes, one can further mention the following other types of civil engineering works that have to be designed against fatigue:

- Sign supporting and traffic sign supporting structures. For these structures, fatigue design can be done similarly to masts. However, there are four wind-loading phenomena which can lead to vibration and fatigue: vortex shedding, galloping, natural wind gusts and, for traffic sign supporting structures, truck-induced gusts. Recommendations have been published in particular in the USA (NCHRP, 2002).
- Offshore platforms. The different aspects of fatigue have been long studied for these structures: action, actions effects, stress computation, fatigue strength, inspection, repair and strengthening, reliability. Thus, there are specific codes for these structures, for example API (2005), DNV (2010) and DNV (2010.2).
- Piers and wharfs. As in offshore platforms and piles, there is often a combination of fatigue and corrosion, which reduces fatigue strength. However, to the authors knowledge, no codes specifically addresses fatigue design for this type of structure.

In all these other types of civil engineering works, the approach to fatigue design and verifications are similar to the ones presented in this book.

3.2 DAMAGE EQUIVALENT FACTORS

3.2.1 Concept

The general concept for the damage equivalent factor, λ, is explained in sub-chapter 1.2. It allows for taking into account real traffic effects (i.e.

3.2 DAMAGE EQUIVALENT FACTORS

fatigue damage effect) by correcting the code load model. Equation (1.5) can thus be rewritten as follows:

$$\gamma_{Ff} \Delta\sigma_{E,2} = \lambda \cdot \Delta\sigma\left(\gamma_{Ff} Q_k\right) \qquad (3.4)$$

The λ value depends on the traffic composition and volume, the design working life, the fatigue load and the static system. The λ value may also depend on the S-N curve, that is on its slope m. Consequently, in the Eurocodes, the general procedure splits the factor λ into 4 different partial factors in order to allow for more parameters to be accounted for, as given in equation (3.5).

$$\lambda = \lambda_1 \cdot \lambda_2 \cdot \lambda_3 \cdot \lambda_4 \text{ but } \lambda \leq \lambda_{max} \qquad (3.5)$$

where
- λ_1 factor accounting for the span length (a relation which is function of the influence line length, see section 3.2.2), function of the structure type, see relevant sections. For example, for road and rail bridges, see section 3.2.3 and 3.2.4 respectively;
- λ_2 factor accounting for the traffic volume, function of the structure type, see relevant sections, for example for road and rail bridges, see section 3.2.3 and 3.2.4 respectively;
- λ_3 factor accounting for the design working life of the structure (e.g. the working life of 100 years for a bridge leads to $\lambda_3 = 1$). For another design working life, the following simplified equation (3.6) applies:

$$\lambda_3 = \left(\frac{t_{Ld}}{100}\right)^{1/m} \qquad (3.6)$$

where t_{Ld} is the design working life of the structure in years and m is the slope of the S-N curve.
- λ_4 factor accounting for the influence of more than one load on the structural member, function of structure type and definition of the fatigue load model for computing λ_1. For example for road and rail bridges, see section 3.2.3 and 3.2.4 respectively;
- λ_{max} maximum damage equivalent factor value, taking into account the fatigue limit. Again, it is a function of the structure type, see relevant sections. For example, for road and rail bridges, see section 3.2.3 and 3.2.4 respectively.

3. DETERMINATION OF STRESSES AND STRESS RANGES

Note that for crane supporting structures, this general procedure is not followed because of the definition of the fatigue load models. In particular, the factor λ_4 is not given, i.e. the effect of more than one load on the crane beam runway is defined differently, see section 3.2.5.

In Figure 3.5, the damage equivalent factor λ_1 for road and rail bridges is represented as a function of the span length, L. The λ_1 values depend on the closeness of the fatigue load model to real loadings, thus a direct comparison between the different curves cannot be made. However, it can be seen that the correction is much more significant for road bridges ($\lambda_1 \approx 2.0$) than for railway bridges ($\lambda_1 \approx 1.0$). This means that for road bridges, the load model 3 deviates substantially from the real traffic volume and loads whereas for railway bridges, the fatigue load model 71 approach corresponds to the operating traffic loads. For road bridges, to achieve the damage equivalence, a clearly higher partial factor λ_1 is necessary. Figure 3.5 also illustrates that for road bridges there is a need for two different curves depending on the position of the detail on the bridge (mid-span or support).

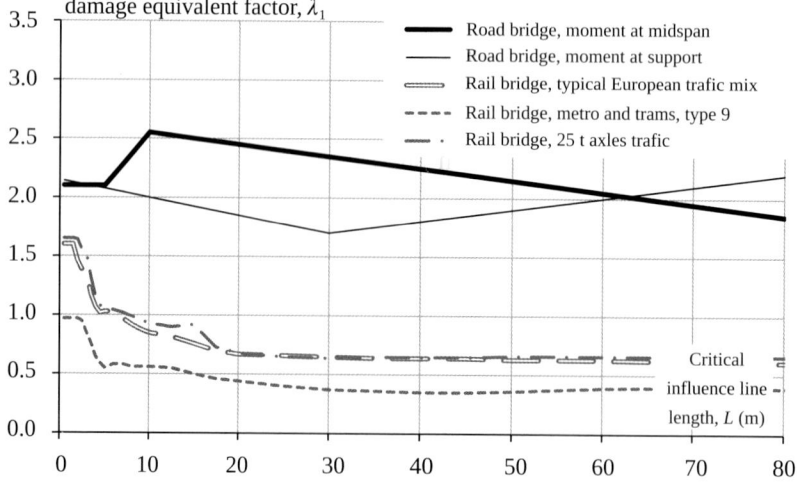

Figure 3.5 – Damage equivalent partial factor λ_1 for road and rail bridges as a function of the critical influence line length L (Nussbaumer, 2006)

The limiting maximum damage equivalent factor, λ_{max}, is dictated by the fact that the multiplication of the individual partial factor may result in a value far exceeding the one obtained from a design using the fatigue limit. Again, there is a significant difference between road and railway bridges.

3.2 DAMAGE EQUIVALENT FACTORS

In the case of railway bridges, the load model represents an upper bound value in terms of the maximum stress range it generates. Thus, the limiting value is bound by the CAFL value, $\Delta\sigma_D$. Before being rounded up to $\lambda_{max} = 1.4$ as in EN 1993-2, it can be expressed as (to express the fact that verification is carried out at 2 million cycles):

$$\lambda_{max} = \frac{\Delta\sigma_C}{\Delta\sigma_D} = \left(\frac{5}{2}\right)^{1/3} = 1.36 \qquad (3.7)$$

In the case of road bridges, it cannot be expressed as a single value since the load model does not represent an upper bound value in terms of the maximum stress range it generates. Therefore, as for the damage equivalent factor λ_1, simulations have been carried out to define proper λ_{max} values (Sedlacek and Müller, 2000). In EN1993-2, it results in values for λ_{max} comprised between 1.8 and 2.7, depending on the bridge span as for λ_1. Expressions for λ_{max} are given in equations (3.13) to (3.15).

3.2.2 Critical influence line length

The determination of the partial damage equivalent factors requires the value of the so-called critical influence line length (or also length of critical influence area) of the cross section and internal force under consideration. Eurocode 3, part 2, gives the following rules for determining it:

a) for bending moments:
- for a simply supported beam, its span length;
- for continuous beams in mid-span regions, see Figure 3.6, the span length L_i of the span under consideration;
- for continuous beams in support regions, see Figure 3.6, the average of the two spans L_i and L_j adjacent to that support;
- for cross-girders supporting stringers, or rail bearers, the sum of the two adjacent spans of the stringers carried by the cross-girder
- for a deck plate supported only by cross-girders or cross-ribs (no longitudinal members), the length of the influence line relevant to compute the deck plate deflection, ignoring any part indicating upward deflection. The same applies for the supporting cross-members. In railway bridges, the stiffness of the rails in the load distribution is to be accounted for.

3. DETERMINATION OF STRESSES AND STRESS RANGES

b) for shear, for both simply supported and continuous beams:
   - for intermediate support regions, see Figure 3.6, the span under consideration L_i;
   - for mid-span regions, see Figure 3.6, $0.4 \cdot L_i$. L_i being the span under consideration.
c) for support reactions:
   - for abutments (i.e. end supports), the span under consideration L_i;
   - for intermediate supports, the sum of the two adjacent spans $L_i + L_j$.
d) for arch bridges:
   - for hangers, twice the length of hangers;
   - for arch, half the span of the arch.

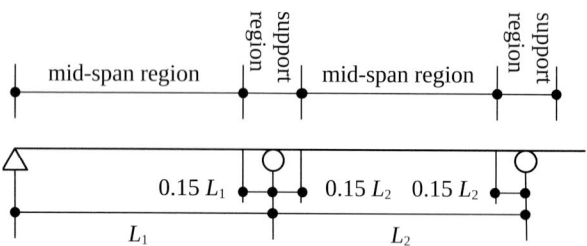

Figure 3.6 – Definition of mid-span and intermediate support regions

3.2.3 Road bridges

In EN 1993-2, λ_1 is only defined for fatigue details subject to stresses due to the global bending moment in road bridges and for span lengths between 10 m and 80 m. There are recommendations to extrapolate using the very same formulae if the span length exceeds 80 m, but they may not be conservative and using a constant value above 80 m may be more realistic (both for mid-span and intermediate support regions).

As previously shown on Figure 3.6, a distinction is also made between the details located in an intermediate support region (15 % of the adjacent span lengths) and those located in a mid-span region.

- In a mid-span region with a span length L (in meters):

$$\lambda_1 = 2.55 - 0.7 \cdot \frac{L-10}{70} \qquad (3.8)$$

3.2 DAMAGE EQUIVALENT FACTORS

- In an intermediate support region between a span length L_1 and a span length L_2, both in meters (see Figure 3.6):

$$\lambda_1 = 2.0 - 0.3 \cdot \frac{L-10}{20} \quad \text{for } 10\text{m} \leq L = \frac{L_1+L_2}{2} \leq 30\text{m} \tag{3.9}$$

$$\lambda_1 = 1.70 + 0.5 \cdot \frac{L-30}{50} \quad \text{for } 30\text{m} \leq L = \frac{L_1+L_2}{2} \ (\leq 80\text{m}) \tag{3.10}$$

- For the abutments, the λ_1 value in the end spans applies.

For shear studs, since the fatigue strength curve slope coefficient is much higher than for other details ($m = 8$ instead of 3 and/or 5, see section 4.1.1), the damage equivalent factor, called $\lambda_{v,1}$, is dealt with separately. This is explained in EN 1994-2 (as well as in EN 1994-1-1, 6.8.6.2). The value for road bridges is a constant, $\lambda_{v,1} = 1.55$.

For the λ_2 factor, the reference value (equal to 1.0) is defined for a traffic in which all the lorries have the same weight ($Q_0 = 480$ kN, the same weight as the Fatigue Load Model no 3 from EN 1991-2) and which contains $N_0 = 0.5 \cdot 10^6$ lorries per year and per slow lane defined on the bridge deck. For a different traffic, the following simplified equation is proposed in EN 1993-2:

$$\lambda_2 = \frac{Q_{ml}}{Q_0}\left(\frac{N_{obs}}{N_0}\right)^{1/m} \tag{3.11}$$

where
- Q_{ml} mean weight (in kN) of the heavy load vehicles on the slow lane according to the formula $(\Sigma n_i Q_i^m / \Sigma n_i)^{1/m}$,
- N_{obs} the number of heavy vehicles which can be expected per year in the slow lane,
- m S-N curve slope; 2nd slope in the case of a S-N curve with two slopes (largest m value).

The λ_3 factor accounts for the design life of the structure. It is obtained from equation (3.6) in section 3.2.1.

If the bridge has more than one slow lane, the value of the λ_4 factor has also to be computed. Since the fatigue load model FLM3 is a single vehicle circulating on one lane and λ_1 is defined with respect to the number of heavy vehicles per slow lane, then λ_4 accounts for the effects of heavy

3. Determination of Stresses and Stress Ranges

vehicles on the other lanes and its value is thus always greater than unity. EN 1993-2 suggests the following equation for λ_4:

$$\lambda_4 = \left[1 + \sum_{j=2}^{k} \frac{N_j}{N_1} \left(\frac{\eta_j Q_{mj}}{\eta_1 Q_{ml}} \right)^m \right]^{1/m} \qquad (3.12)$$

where

- N_1, N_j number of lorries per year in slow lane 1, respectively lane j, see indications in Table 3.3,
- Q_{m1}, Q_{mj} average gross weight of the lorries in the slow lane 1 and lane j respectively,
- k total number of slow lanes,
- m S-N curve slope; 2nd slope in the case of a S-N curve with two slopes (largest m value),
- η_1, η_j value of the influence line for the internal force that produces the stress range in the middle of the lane j (with positive sign).

Note that the influence of the percentage of crossings or overtakings is not considered (i.e., neglected in the above equation).

For the λ_{max} factor, as for the λ_1 factor, it is only defined for fatigue details subject to stresses due to the global bending moment in road bridges and for span lengths between 10 m and, strictly, 80 m. Expressions for the λ_{max} factor are as follows (see Figure 3.6 for region definition):

- In a mid-span region with a span length L (in meters):

$$\lambda_{max} = 2.5 - 0.5 \frac{L-10}{15} \geq 2.0 \qquad (3.13)$$

- In an intermediate support region between a span length L_1 and a span length L_2, both in meters:

$$\lambda_{max} = 1.8 \text{ for } L = \frac{L_1 + L_2}{2} \leq 30\,\text{m} \qquad (3.14)$$

$$\lambda_{max} = 1.80 + 0.9 \cdot \frac{L-30}{50} \text{ for } 30\,\text{m} \leq L = \frac{L_1 + L_2}{2} \ (\leq 80\,\text{m}) \qquad (3.15)$$

- For the abutments, the λ_{max} value used for end spans applies.

Finally, for shear studs, $\lambda_{v,2}$, $\lambda_{v,3}$, $\lambda_{v,4}$ factors should be determined using the relevant equations, but using a slope coefficient $m = 8$, or exponent

$1/8$, in place of those given to allow for the relevant fatigue strength curve for headed studs in shear, see section 4.1.1. Regarding λ_{max}, the conservative values given in equations (3.13) to (3.15) may be used.

Example 3.2: Application to steel and concrete composite road bridge, determination of the partial damage equivalent factors (worked example 1).

The side spans of the road bridge studied are 90 m long and the main spans are 120 m long, so that the formulae (3.8) to (3.10) for the λ_1 factor can still be used as approximations. This leads to the Table 3.7 for a structural detail submitted to a stress range resulting from an applied bending moment.

Table 3.8 – Partial damage equivalent factor λ_1 value for road bridge detail

Section location	L (m)	λ_1
In end span	90	$2.55 - 0.7 (L-10)/70 = 1.75$
Around P1, P4 supports	$(90+120)/2 = 105$	$1.7 + 0.5 (L-30)/50 = 2.45$
Around P2, P3 supports	120	$1.7 + 0.5 (L-30)/50 = 2.60$
In central spans	120	$2.55 - 0.7 (L-10)/70 = 1.45$

The λ_2 factor must be used since the traffic is different from the reference traffic ($Q_0 = 480$ kN, $N_0 = 0.5 \cdot 10^6$ lorries). The resulting value is $\lambda_2 = 1.223$ and its determination is detailed in Example 3.3. Note that if it is usual to design with different number of lorries, N_{obs}, it is less common to have information about weight distribution of the lorries, Q_{ml} from the authority.

The λ_3 factor follows the design working life of the structure. For a bridge this lifetime is generally equal to 100 years so that $\lambda_3 = 1.0$.

The bridge example is a box-girder bridge with a large highway slab on which two slow lanes have to be considered (one in each direction). Assuming that each slow lane supports the same traffic, it means $Q_{m2} = Q_{m1}$ and $N_2 = N_1$. For a box girder bridge, each slow lane has also the same transverse influence in a structural detail, so $\eta_2 = \eta_1$, and finally, assuming a S-N curve slope $m = 5$, one gets using equation (3.12):

$$\lambda_4 = \left[1 + \frac{N_2}{N_1}\left(\frac{\eta_2 \cdot Q_{m2}}{\eta_1 \cdot Q_{m1}}\right)^5\right]^{1/5} = 1.15$$

Finally, the product of the partial damage equivalent factors must be less than or

3. DETERMINATION OF STRESSES AND STRESS RANGES

equal to λ_{max}. As for the λ_1 factor, the formulae given for λ_{max} in EN 1993-2 can be used as approximation with regards to the span lengths used in the example. For a detail submitted to a bending moment, this leads to Table 3.9.

Table 3.9 – Partial damage equivalent factor λ_{max} value for road bridge detail

Section location	L (m)	λ_{max}
In end span	90	2.0
Around P1, P4 supports	(90+120)/2 = 105	1.8 + 0.9 (L–30)/50 = 3.15
Around P2, P3 supports	120	1.8 + 0.9 (L–30)/50 = 3.42
In central spans	120	2.0

Finally, the damage equivalent factor λ can be computed using expression (3.5), repeated below:

$$\lambda = \lambda_1 \cdot \lambda_2 \cdot \lambda_3 \cdot \lambda_4 \text{ but } \lambda \leq \lambda_{max}$$

Table 3.10 summarizes the results obtained for the whole bridge. In this example, it can be seen that λ_{max} governs.

Note: for a traffic with only $0.5 \cdot 10^6$ lorries per year and direction, λ would have governed.

Table 3.10 – Summary of damage equivalent factor values for road bridge example

Section location (between 0 m and 540 m)	λ
End spans C0-P1 and P4-C5 : - between 0 m and $0.85 \times L_1 = 76.5$ m - between 463.5 m and 540 m	2.66 ($\leq \lambda_{max} = 2.0$) thus 2.0
Supports P1 and P4 : - between 76.5 m and $L_1 + 0.15 \times L_2 = 108$ m - between 432 m and 463.5 m	3.72 ($\leq \lambda_{max} = 3.15$) thus 3.15
Supports P2 and P3 : - between 192 m and 228 m - between 312 m and 348 m	3.95 ($\leq \lambda_{max} = 3.42$) thus 3.42
Central spans P1-P2, P2-P3 and P3-P4 : - between 108 m and 192 m - between 228 m and 312 m - between 348 m and 432 m	2.20 ($\leq \lambda_{max} = 2.0$) thus 2.0

3.2 DAMAGE EQUIVALENT FACTORS

Example 3.3: Determination of λ_2 for a road bridge traffic case not explicitly given in EN 1991-1-2

Note: This example is not necessary for the steel and concrete composite road bridge example. It is an additional example based on the bridge example to explain the possibilities of the partial damage equivalent factors to solve the fatigue design of a bridge for traffic cases (service life, volume) not explicitly given in the codes.

As a start, one uses equation (3.11), repeated below, assuming a S-N curve slope $m = 5$:

$$\lambda_2 = \frac{Q_{ml}}{Q_0} \left(\frac{N_{obs}}{N_0} \right)^{1/5}$$

with $Q_0 = 480$ kN and $N_0 = 0.5 \cdot 10^6$ lorries per year and per slow lane.

N_{obs} and Q_{ml} quantify the real traffic expected on the bridge. Assumptions and/or measurements have to be made by the designer.

For this example, the following assumptions have been made:

- $N_{obs} = 2\ 000\ 000$ lorries per year and per slow lane,
- an average gross weight of the mean lorry deduced from measurements using $Q_{ml} = \left(\sum_i n_i Q_i^5 / \sum_i n_i \right)^{1/5}$ and equal to 445 kN.

According to these assumptions: $\lambda_2 = 1.223$.

The damage equivalent factor λ can now be computed using expression (3.5), repeated below :

$$\lambda = \lambda_1 \cdot \lambda_2 \cdot \lambda_3 \cdot \lambda_4 \text{ but } \lambda \leq \lambda_{max}$$

with λ_1, λ_3, λ_4 and λ_{max} computed in the same way as presented in Example 3.2.

Note: This part is not necessary for the usual design of a steel and concrete composite road bridge. It is an addition made to explain the possibility of the partial damage equivalent factors to solve the fatigue design of a bridge for traffic cases (service life, volume) not explicity given in the codes.

3. DETERMINATION OF STRESSES AND STRESS RANGES

3.2.4 Railway bridges

In the same way as for road bridges, EN 1993-2 defines the damage equivalent factor for steel structural details in a railway bridge with span length up to 100 m. For longer spans, the values at 100 m may be used. The difference with road bridges is that λ_1 is computed with a fatigue load model placed on each of the tracks. Figure 3.7 illustrates the values adopted by EN 1991-2 for the λ_1 factor. These values are smaller than 1.0 because they will be applied to the characteristic LM71 (or SW/0) traffic loads which have been proposed initially for the ultimate limit strength design.

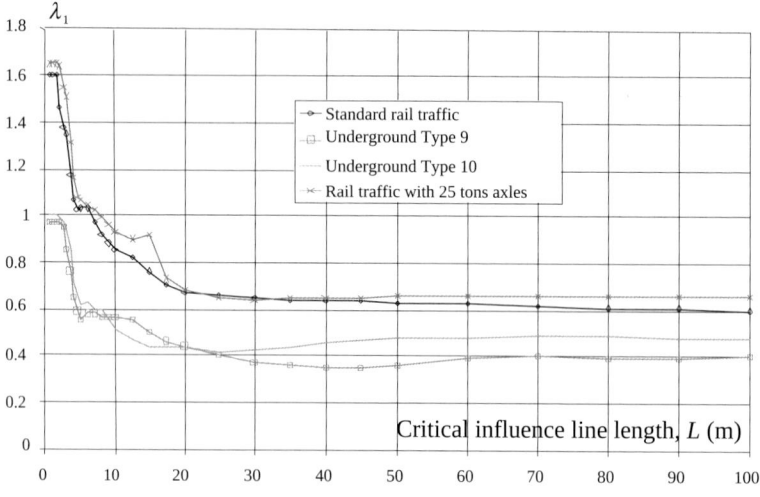

Figure 3.7 – λ_1 factor for railway bridges

As for the influence of the traffic volume, Figure 3.8 shows the relationship between the traffic volume per year and the λ_2 factor values. It can be checked that the reference case, (i.e. where λ_2 is equal to unity and thus disappears in the verification), corresponds to a traffic volume of 25 million tons per year and per track.

For railway bridges with more than one track, since the fatigue load model has to be placed on each of the tracks, λ_1 is defined with respect to the whole bridge traffic volume (same train traffic on each track). Thus, λ_4 accounts for the effects of not having always trains at the same time on the bridge and its value is thus always lower than unity. In other words, the design of a railway bridge with two tracks, including fatigue, is done with the two tracks loaded simultaneously to get the maximum possible stresses and stress range $\Delta\sigma_{1+2}$ (in opposition to what is usually done in the design of road bridges).

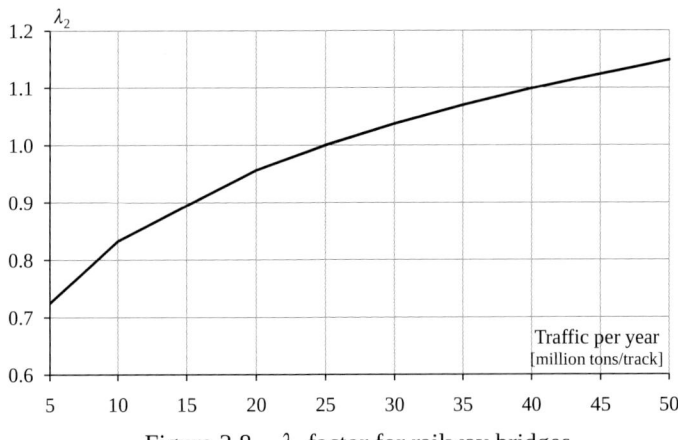

Figure 3.8 – λ_2 factor for railway bridges

The code deals only with bridges with two tracks (or more but not loaded at the same time). The expression for factor λ_4 is the following:

$$\lambda_4 = \left[p + (1+p)\left(a^5 + (1-a)^5\right) \right]^{1/5} \qquad (3.16)$$

where
- $a = \Delta\sigma_1/\Delta\sigma_{1+2}$ ratio between one and two tracks loaded,
- $\Delta\sigma_1$ stress range in the structural detail created by the LM71 train on the track 1,
- $\Delta\sigma_{1+2}$ stress range in the same detail created by the LM71 train on the two considered tracks,
- p percentage of the total traffic that meets on the bridge (percentage of crossings).

The parameter a takes the transverse distribution of the two tracks on the bridge into account. As for the parameter p, EN 1993-2 gives the indication to use a crossing percentage of 12 %.

As already explained in section 3.2.1, the λ_{max} factor for railway bridges is rounded up to 1.4.

For shear studs, as for road bridges, since the fatigue strength curve slope coefficient is much higher than for other details ($m = 8$ instead of 3 and/or 5, see

3. Determination of Stresses and Stress Ranges

section 4.1.1), the damage equivalent factor $\lambda_{v,1}$ had to be computed specifically and is given in EN 1994-2. The value is not a constant as for road bridges, it can be determined from Figure 3.9 (extract from EN 1994-2). The $\lambda_{v,2}$, $\lambda_{v,3}$, $\lambda_{v,4}$ factors should be determined using the relevant expressions, but again using a slope coefficient $m = 8$, or exponent $1/8$, instead of those given. Regarding λ_{max}, it does not exist since there is no CAFL nor cut-off limit.

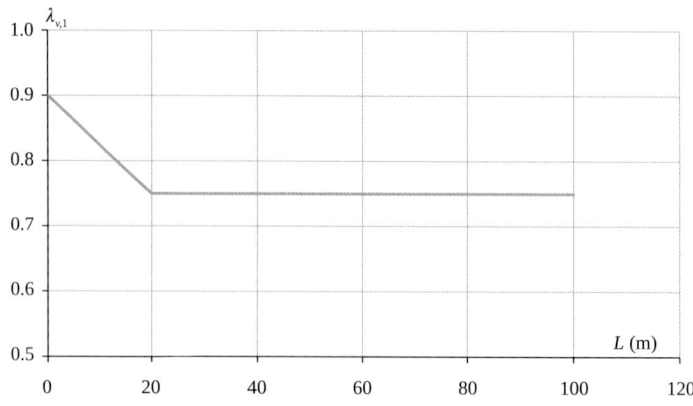

Figure 3.9 – $\lambda_{v,1}$ factor for shear studs in railway bridges

3.2.5 Crane supporting structures

The damage equivalent factors indicated in EN 1991-3 for crane supporting structures are based on standardized stress ranges with a Gaussian distribution of the load effects. The complete spectrum can be simplified in a number of levels, at least 8 like shown in Figure 3.10. Standardized load spectrums are most often described using the following relationship (Radaj et al, 2003), given here for load ranges:

$$N_i = N_{tot}^v \quad v = 1 - \left(\frac{\Delta F_i}{\Delta F_{max}}\right)^n \tag{3.17}$$

where
- N_i number of cycles exceeding load range ΔF_i,
- N_{tot} total number of cycles in the spectrum,
- ΔF_i load range of level i,
- ΔF_{max} maximum load range in the spectrum,
- n exponent defining the exceedance rate curve shape; $n = 2$ corresponds to a Gaussian distribution.

3.2 DAMAGE EQUIVALENT FACTORS

The linear accumulation of damage is done according to *Miner's rule* and the fatigue strength curves contained in EN 1993-1-9 with a constant slope of $m = 3$ for direct stress ranges and $m = 5$ for shear stress ranges.

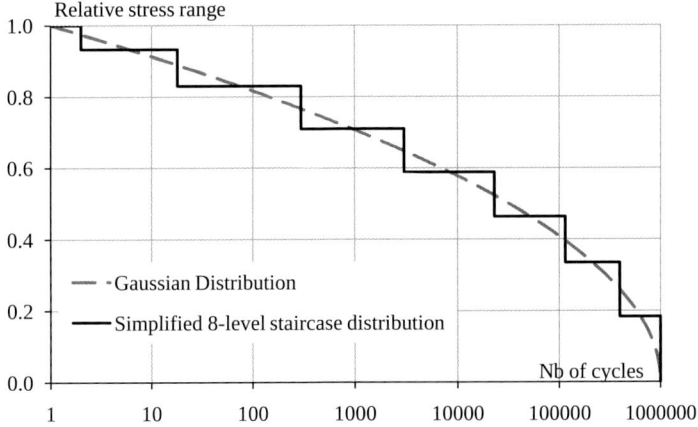

Figure 3.10 – Gaussian distribution standardised as relative stress range for crane supporting structures (tracks) (IIW, 2009)

In computing the fatigue stress spectra, the secondary moments due to joint rigidity and chord member continuity in members of lattice girders, lattice surge girders and triangulated bracing panels should be included.

The secondary moments in triangulated structural members may be computed using an adequate model of the triangulated structure, which is modelling properly the behaviour and stiffness of the members and joints. Since this is a difficult task, correction factors can also be used on the primary stresses. In EN 1993-6, three different cases are considered, namely:

- members of lattice girders, lattice surge girders and triangulated bracing panels. In this case, secondary moments due to joint rigidity may be considered by using k_1-factors as specified in clause 4(2) of EN 1993-1-9 (tables 4.1 and 4.2, originally developed for hollow section joints, see section 3.3.6).
- members of open cross section. In this case, the k_1-factors given in Table 3.11 (extracted from EN 1993-6, Table 5.4) may be used. In this table, the distance y represents the distance between the potential crack position at edge and the centroid of the member.
- members made from structural hollow sections with welded joints. In this case, the k_1-factors given in tables 4.1 and 4.2 of EN 1993-1-9 may be used.

3. DETERMINATION OF STRESSES AND STRESS RANGES

Table 3.11 – Coefficients k_1 for secondary stresses in members of open sections
(source Table 5.4 from EN 1993-6)

Lattice girders loaded only at nodes			
Range of L/y values	$L/y \leq 20$	$20 \leq L/y < 50$	$L/y \geq 50$
Chord members End and internal members	1.57	$\dfrac{1.1}{0.5 + 0.01\, L/y}$	1.1
Secondary members*	1.35	1.35	1.35
Lattice girders with chord members loaded between nodes			
Range of L/y values	$L/y < 15$		$L/y \geq 15$
Loaded chord members	$\dfrac{0.4}{0.25 + 0.01\, L/y}$		1.0
Unloaded chord members Secondary members*	1.35		1.35
End members	2.5		2.5
Internal members	1.65		1.65

L is the length of the member between nodes;
y is the perpendicular distance, in the plane of triangulation, from the centroid axis of the member to its relevant edge, measured as follows:
- compression chord: in the direction from which the loads are applied
- tension chord: in the direction in which the loads are applied
- other members: the larger distance.

* Secondary members comprise members provided to reduce the buckling lengths of other members or to transmit applied loads to nodes. In an analysis assuming hinged joints, the forces in secondary members are not affected by loads applied at other nodes, but in practice they are affected due to joint rigidity and the continuity of the chord members at joints.

The damage equivalent factor λ_i is the correction factor to make allowance for the relevant standardized fatigue load spectrum and absolute number of load cycles in relation to $N = 2 \cdot 10^6$ cycles.

$$Q_{E,2} = \varphi_{fat} \cdot \lambda_i \cdot Q_{r,\max,i} \tag{3.18}$$

where

$Q_{r,\max,i}$ maximum value of the characteristic vertical wheel load i
$\lambda_i = \lambda_{1,i} \cdot \lambda_{2,i}$

φ_{fat} damage equivalent dynamic impact factor (EN 1991-3, clause 2.12.1(7))

3.2 Damage Equivalent Factors

$\lambda_{1,i}$ damage equivalent factor to make allowance for the relevant standardised fatigue load spectrum

$\lambda_{2,i}$ damage equivalent factor to translate the number of cycles at 2 millions

$$\lambda_{1,i} = \sqrt[m]{kQ} = \left[\sum_j \left(\left(\frac{\Delta Q_{i,j}}{\max \Delta Q_i} \right)^m \frac{n_{i,j}}{\sum n_{i,j}} \right) \right]^{1/m} \tag{3.19}$$

$$\lambda_{2,i} = \sqrt[m]{nv} = \left[\frac{\sum_j n_{i,j}}{N} \right]^{1/m} \tag{3.20}$$

Alternatively, λ_i can be found in EN 1991-3, Table 2.12 (reproduced in Table 3.12 below) after classification of the crane according to the load spectrum and the total number of load cycles which can be made following EN1991-3, Table 2.11 (reproduced in Table 3.6).

Table 3.12 – λ_i-values according to the classification of cranes (source Table 2.12 from EN 1991-3)

Classes	S_0	S_1	S_2	S_3	S_4	S_5	S_6	S_7	S_8	S_9
Normal stresses	0.198	0.250	0.315	0.397	0.500	0.630	0.794	1.000	1.260	1.587
Shear stresses	0.379	0.436	0.500	0.575	0.660	0.758	0.871	1.000	1.149	1.320
Note 1 : In determining the λ_i-values, standardized spectra with a Gaussian distribution of the load effects, the Miner rule and fatigue strength S-N lines with a slope $m = 3$ for normal stresses and $m = 5$ for shear stresses have been used.										
Note 2: In case the crane classification is not included in the specification documents of the crane indications are given in EN 1991-3, annex B.										

Simplified rules exist for crane supporting structures, in particular rail girders, supporting more than one crane and thus subjected to combined action effects from those cranes. These rules are similar to those for example for bridges, making use of the partial damage equivalent factor λ_4. In the absence of better information, it is recommended to take a value of λ_4, which is called λ_{dup} in this case, equal to the values λ_i from Table 3.12 for a loading class S_i as follows:

3. Determination of Stresses and Stress Ranges

- for 2 cranes: 2 classes below the loading class of the crane with lower loading class;
- for 3 or more cranes: 3 classes below the loading class of the crane with the lowest loading class.

This leads to a slightly different expression (when compared to expression (3.18)), that is expression (3.21) for the equivalent characteristic vertical load $Q_{E,2,dup}$ due to two or more cranes occasionally acting together.

$$Q_{E,2,dup} = \varphi_{fat} \cdot \lambda_{dup} \cdot Q_{r,max,dup} \qquad (3.21)$$

where

$Q_{r,max,dup}$ maximum value of the characteristic vertical wheel loads from all cranes acting together,

λ_{dup} partial damage equivalent factor for the effects of all cranes acting together.

The stress ranges are deduced from the vertical loads. Verification can use stress ranges or the damage accumulation, see sub-chapter 5.4.

Example 3.4: Application to runway beam of crane (worked example 3)

In this example, the fatigue loads and stresses of the runway beam of the crane are determined. The general description, geometry and dimensions are given in section 1.4.4.

The crane is classified as Q4 (class of load spectrum) and U4 (class of total number of cycles) – corresponding to fatigue actions class S3 according to Table 2.11 of EN 1991-3 (see Table 3.6) and hoisting class HC4 (see Recommendations in Annex B of EN 1991-3)

Fatigue load:

The equivalent fatigue load $Q_{E,2}$ is calculated using equation (3.18).

$$Q_{E,2} = \varphi_{fat} \cdot \lambda_i \cdot Q_{r,max,i}$$

where the damage equivalent factor value from Table 3.12 is $\lambda_i = 0.397$, according to the crane classification (S3).

3.2 Damage Equivalent Factors

The damage equivalent dynamic impact factor φ_{fat} for normal conditions may be taken, if not specified by the supplier, as the maximum of the following two values (EN 1991-3, § 2.12.1 (7)):

$$\varphi_{fat,1} = \frac{1+\varphi_1}{2} = 1.05$$

$\varphi_{fat,1}$ is a factor accounting for the excitation of the crane structure due to lifting the hoist load off the ground. For overhead travelling bridge cranes $\varphi_1 = 1.1$.
$0.9 < \varphi_1 < 1.1$ - The two values 1.1 and 0.9 reflect the upper and lower values of the vibrational pulses.

$$\text{and } \varphi_{fat,2} = \frac{1+\varphi_2}{2}$$

$\varphi_{fat,2}$ is a factor accounting for the dynamic effects of transferring the hoist load from the ground to the crane.

For computing the dynamic effects of transferring the hoist load, the following parameters are needed:

$$\varphi_2 = \varphi_{2,\min} + \beta_2 \cdot v_h$$

v_h steady hoisting speed in m/s (specified by the supplier). A steady hoisting speed of $v_h = 0.2$ m/s is assumed

$\varphi_{2,\min}$ and β_2 from EN 1991-3, Table 2.5 (reproduced in Table 3.13 below).

Table 3.13 – Values of β_2 and $\varphi_{2,\min}$ (EN 1991-3, Table 2.5)

Hoisting class of appliance	β_2	$\varphi_{2,\min}$
HC1	0.17	1.05
HC2	0.34	1.10
HC3	0.51	1.15
HC4	0.68	1.20

For HC4, $\beta_2 = 0.68$ and $\varphi_{2,\min} = 1.2$

$$\varphi_2 = 1.2 + 0.68 \cdot 0.2 = 1.336$$

$$\varphi_{fat,2} = \frac{1+1.336}{2} = 1.168$$

$$\varphi_{fat} = \max\left(\varphi_{fat,1}, \varphi_{fat,2}\right) = \max(1.05, 1.168) = 1.168$$

3. DETERMINATION OF STRESSES AND STRESS RANGES

$\varphi_{fat,2} = 1.168$ and thus to $\varphi_{fat} = 1.168$

The equivalent fatigue load, per wheel, is then equal to:

$$Q_{E,2} = 1.168 \cdot 0.397 \cdot 73.4 = 34.4 \text{ kN}$$

Stress ranges:

In general, the difference of bending moments can be expressed as (see relevant crane calculation tables):

$$\Delta M(Q_{E,2}) = M_{max}(Q_{E,2}) - M_{min}(Q_{E,2}) = M_{max}(Q_{E,2}) \cdot \left(1 - \frac{M_{min}(Q_{E,2})}{M_{max}(Q_{E,2})}\right)$$

The stress ranges can then be calculated as

$$\Delta \sigma_{E,2} = \frac{\Delta M(Q_{E,2})}{I_y} \cdot z$$

with $I_y = 155.8 \cdot 10^6 \text{ mm}^4$

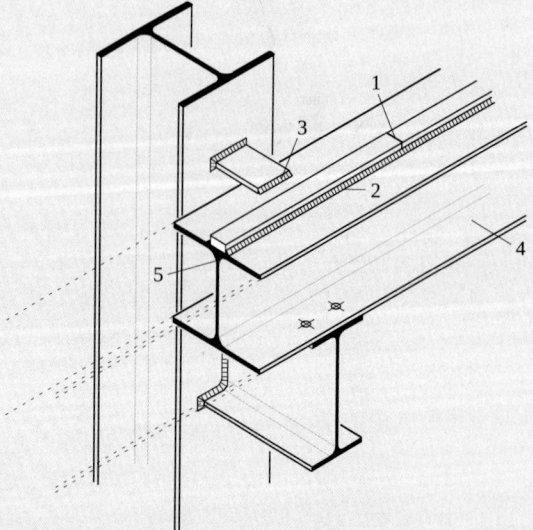

Figure 3.11 – Critical fatigue locations to be verified

Table 3.14 gives a summary of the computed moments and equivalent stress ranges at details critical points (see Figure 3.11). Stress ranges are given with their signs even though the signs may be omitted.

3.2 DAMAGE EQUIVALENT FACTORS

Table 3.14 – Summary of the moments and stress ranges in the runway beam

detail	z [mm]	Support		Mid-span	
		ΔM [kNm]	$\Delta\sigma_{E,2}$ [N/mm²]	ΔM [kNm]	$\Delta\sigma_{E,2}$ [N/mm²]
1	-141	-37.1	33.6	45.5	-41.2
2	-121	-37.1	28.8	45.5	-35.3
3	-121	-37.1	28.8	-	-
4	149	-37.1	-	45.5	43.5
5	-98	-37.1	23.3	45.5	-28.6

(+ for tension and – for compression)

If there were two cranes working together, $Q_{E,2,dup}$ would have to be computed to estimate the additional effects of two or more cranes occasionally acting together (see expression (3.21)). As a simplification it is considered that the two cranes acting together have the same fabricator specifications.

λ_{dup} should be taken from Table 3.12 (using 2 classes below the loading class of the crane), $\lambda_{dup} = 0.250$ and consequently:

$$Q_{E,2,dup} = \varphi_{fat} \cdot \lambda_{dup} \cdot Q_{r,max,dup}$$

$$Q_{E,2,dup} = 1.168 \cdot 0.250 \cdot 2 \cdot Q_{r,max,i} = 42.9\,\text{kN}$$

The same procedure as before is used to compute $\Delta M(Q_{E,2dup})$ and $\Delta\sigma_{E,2dup}$. The results are summarized in Table 3.15.

Table 3.15 – Summary of the additional moments and stress ranges in the runway beam with cranes occasionally acting together

detail	z [mm]	Support		Mid-span	
		ΔM [kNm]	$\Delta\sigma_{E,2dup}$ [N/mm²]	ΔM [kNm]	$\Delta\sigma_{E,2dup}$ [N/mm²]
1	-141	-46.7	42.3	57.3	-51.8
2	121	-46.7	36.3	57.3	-44.5
3	-121	-46.7	36.3	-	-
4	149	-46.7	-	57.3	54.8
5	-98	-37.1	29.4	45.5	-36.0

(+ for tension and – for compression)

3. Determination of Stresses and Stress Ranges

3.2.6 Towers, masts and chimneys

In the case of towers, masts and chimneys, the damage equivalent factor λ serves only to transform the applied stress range $\Delta\sigma_E$ associated to N cycles into an applied stress range $\Delta\sigma_{E,2}$ associated to 2 million cycles. Thus, the damage equivalent factor λ is equivalent to the partial damage equivalent factor λ_1 and may be determined as follows:

$$\lambda_1 = \left(\frac{N}{2 \cdot 10^6}\right)^{1/m} \tag{3.22}$$

where m is the slope of the relevant S-N curve.

In case of a change in the design life of the structure, expression (3.22) can be applied without the need for another partial damage equivalent factor. Since wind loadings often cause a very large number of cycles over the service life, instead of the above, a fatigue design using the fatigue limit is often necessary, see section 5.4.2. In this case, if the verification is carried out using directly the fatigue strength expressed as $\Delta\sigma_D$, no damage equivalence factor is required $\lambda_1 = \lambda = 1.0$.

Example 3.5: Computation of total number of cycles, application to chimney (worked example 2)

Number of cycles during the design life

Using expression (3.3), the design value of the number of load cycles (50 years), N_v, is:

$$N_v = 2 \cdot T \cdot n_{y,1} \cdot \varepsilon_0 \cdot \left(\frac{v_{crit,1}}{v_0}\right)^2 \cdot e^{\left(-\left(\frac{v_{crit,1}}{v_0}\right)^2\right)} = 8.6 \cdot 10^8 \text{ cycles}$$

With:

  Service life $T = 3.2 \cdot 10^7 \cdot 50$ seconds
  Bandwidth factor, conservative value taken as $\varepsilon_0 = 0.3$
  Critical wind velocity 1st mode $v_{crit,1} = 5.304$ m/s
  Velocity linked with modal value $v_0 = 0.2 \cdot 30.8 = 6.16$ m/s
  Fundamental bending frequency, 1st mode $n_{y,1} = 0.586$ Hz

Due to the large number of load cycles (> 10^8), the only possible verification to satisfy fatigue design is to do it with the fatigue limit, see section 5.4.2. In other words, this means that one will require stress ranges to stay sufficiently low to have infinite life for all the structural details of the chimney. No damage equivalent factor is needed in this case, i.e. $\lambda_1 = \lambda = 1.0$.

3.3 CALCULATION OF STRESSES

3.3.1 Introduction

In EN 1993-1-9 chapters, the calculation of stresses resulting from the action effects is separated from the calculation of stress ranges, thus the authors chose to respect this separation in this manual. Another advantage of this separation is that one method of calculating stresses may be used in different methods of calculation of stress ranges. Note that dynamic effects are usually considered within the load models (see sub-chapter 3.1) and are not dealt with here. Tensile stresses are considered positive and compressive stresses negative, but this is in general irrelevant since only the ranges of cyclic stress are used in fatigue (i.e. mean stresses are neglected in welded structures). However, the stress signs are of relevance for non-welded details under, nominally, fully or partly compressive loading in computing the stress range (see section 3.7.2).

The procedures for determining the stresses, as well as stress ranges, in the fatigue analysis must agree with the test results analysis made to define the fatigue strength values (i.e. detail categories in the tables of EN 1993-1-9 given as $\Delta\sigma_C$). Thus, even though some methods may not be truly correct from the point of view of the strength of material, consistency is more important than method accuracy. The consequence of the above is that the information given in this section to compute stresses applies directly only together to the details given in the Eurocode tables, not with fatigue strengths taken from other standards, recommendations or references without proper validation. In most situations, the location where a potential fatigue crack will develop is located in the parent material adjacent to some form of stress concentration (e.g. hole, corner, weld). As an example, Figure 3.12 shows on the left a detail from the tables in EN 1993-1-9 (Table 8.5, detail 6, end of coverplate) and on the right a fatigue crack that developed at this detail. But fatigue cracks can also start from heat affected zones (for example Table 8.1, detail 5) or in the weld material (as for example in Table 8.2, detail 10 or Table 8.3, detail 13).

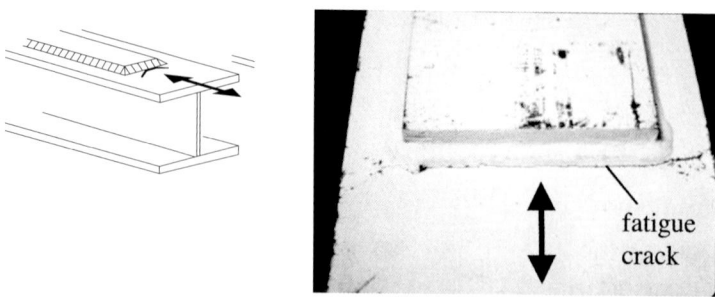

Figure 3.12 – Example of detail (from EN 1993-1-9, Table 8.5, detail 6, end of coverplate) and fatigue crack that developed at this detail

The relevant stress in the member must be calculated in accordance with the arrow, position and direction, indicated for each detail in tables 8.1 to 8.10 of EN 1993-1-9 (which strictly speaking represents the stress range).

Different subdivisions in the methods for computing stresses can be made and the next sub-chapter and sections are dedicated to those methods. The first subdivision in computing stresses, nominal stresses in this case, that can be made is between computing stresses in the parent material (treated in sections 3.3.2, 3.3.5 and 3.3.6) versus computing stresses in bolts (section 3.3.3) or in welds (section 3.3.4). The second subdivision that can be made is between computing nominal stresses versus "modified" stresses, such as modified nominal stresses (sub-chapter 3.4) or geometric stresses (sub-chapter 3.5). Finally, the third and last subdivision is between different types of structural systems, which can lead to some difficulties such as steel and concrete composite sections (section 3.3.5), tubular structures (section 3.3.6 and 3.5.3) or orthotropic decks (sub-chapter 3.6).

3.3.2 Relevant nominal stresses

This is the first and simplest approach. Nominal stresses in parent material should be calculated using elastic theory and accounting for all axial, bending and shearing stresses occurring under the design loading. No redistribution of loads or stresses, such as from plate buckling theory for checking static strength at ultimate limit state, including implicit allowance for redistribution in simplified elastic design rules, or for plastic design procedures, may be allowed. For each detail in tables 8.1 to 8.10 of EN 1993-1-9, unless indicated otherwise, the location and direction where the nominal stress is to be calculated is indicated by the arrow. The plane on which the nominal stress should be calculated is perpendicular to the arrow, and of course parallel to the crack that developed (see the example in Figure 3.12).

The detail categories in the EN 1993-1-9 tables include the effect of local stress concentration due to weld shape, discontinuities, imperfections, triaxiality, residual stresses due to welding, etc. Usually, the relevant nominal stresses under fatigue loading can be computed using the same structural system model as for the static analysis. Unless otherwise noted in tables 8.1 to 8.10, the stress should always be based on the net section. An important point is that the effect for arising stresses in the detail such as:

– eccentricity of global structural axis (e.g. eccentricities in triangulated skeletal structure which introduce secondary moments),
– imposed deformations,
– unintentional movement,
– effective fixity (partial joint stiffness) and
– cracking of concrete in composite members (see section 3.3.5)

should be calculated and taken into account when determining the nominal stress at the detail. The detail should conform exactly to a detail geometry and thus to a category as given in the tables 8.1 to 8.10.

If there are imperfections and macro-geometric features at the detail that change the nominal stress distribution, the structural stress analysis should be refined accordingly. More precisely, the following effects have to be considered:

– local eccentricity,
– misalignment, if the value exceeds the fabrication tolerance which is included in the detail category, etc.,
– stress distribution in the vicinity of concentrated loads,
– shear lag (see Figure 3.13), restrained torsion and distortion, transverse stresses and flange curvature.

3. Determination of Stresses and Stress Ranges

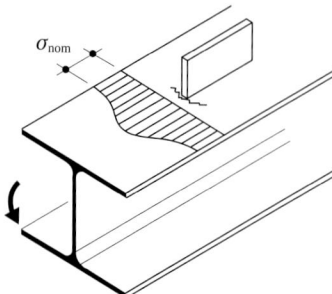

Figure 3.13 – Wide flange beam in bending with a structural detail on its flange

3.3.3 Stresses in bolted joints

There is a distinction, somewhat unclear in EN 1993-1-9, between the detail categories for bolts and the detail categories for bolted connections.

In the case of detail categories for bolted connections (i.e. Table 8.1: details 8 to 13) one has to compute the nominal stresses (and stress ranges) in the elements composing the connection. A further split is made between preloaded and non-preloaded connections. In the case of preloaded connections, logically, the gross cross section area is used to compute the nominal stress; in the case of non-preloaded connections, the net cross section area shall be used to compute the nominal stress (without consideration for the non-uniform stress distribution across the plate). Note that in the case of non-preloaded bolts with normal clearance holes, since the connection may slip by a significant amount, one must ensure that it remains in bearing pressure (contact) all the time and thus no load reversals are permitted. Also the problem of nuts coming loose due to cyclic loading must be addressed.

In the case of the detail category for bolt in tension (i.e. Table 8.1: detail 14), the modified nominal stress is to be used in order to account for the possible increase in stress range due to bending. As an example, bending may result from eccentricities or prying effects (see sub-chapter 3.4 on modified nominal stress and section 3.7.3 on stress range in bolted joints).

3.3.4 Stresses in welds

One can differentiate between load carrying welded joints and the other welded joints such as those in welded build-up sections or transverse butt welds. For the latter, the relevant nominal stresses in the parent material or in the section

3.3 CALCULATION OF STRESSES

at the position of the weld, e.g. for longitudinal welds, shall be calculated (see section 3.3.2). For the former, the calculation of the stresses for fatigue differs from the procedure given for the verification of fillet welds or partial penetration welds for ultimate limit state as given in EN 1993-1-8. The calculation of stresses in the weld for fatigue refers to the projected effective throat section as shown in Figure 3.14. Since tensile residual stresses are assumed to exist in all welded joints, none of the load is carried in bearing between parent materials (in the gap), even if the joint is under compressive loads.

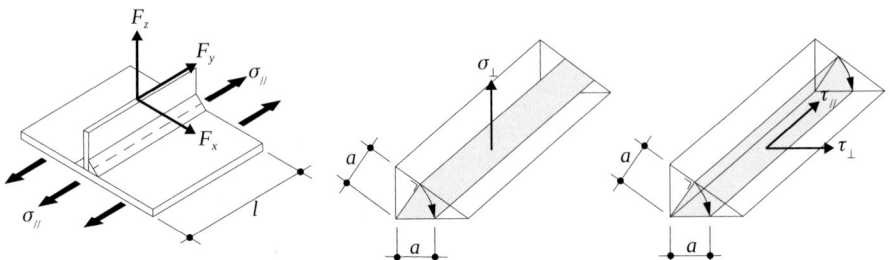

Figure 3.14 – Relevant stresses in fillet welds (double-sided is represented), also valid for partial penetration welds

Depending upon the corresponding detail, the necessary nominal stresses are calculated as given below in expressions (3.23) to (3.25). The force resultants are defined so that the moments caused by F_x and F_y do not need to be considered (i.e. can be neglected). For clarity, the combined size of effective weld throats is used, which corresponds for double-sided fillet or partial penetration welds, as shown in Figure 3.14, to $w = 2a$.

- Nominal resulting normal stress transverse to the axis of the weld:

$$\sigma_w = \sqrt{\sigma_\perp^2 + \tau_\perp^2} \text{ with } \tau_\perp = \frac{F_x}{w \cdot \ell} \qquad (3.23)$$

- Nominal resulting longitudinal shear stress:

$$\tau_w = \tau_\parallel = \frac{F_y}{w \cdot \ell} \qquad (3.24)$$

- Global normal stress (in the load carrying plate):

$$\sigma_\parallel = \frac{F_{y,global}}{A} \qquad (3.25)$$

3. DETERMINATION OF STRESSES AND STRESS RANGES

where

$F_{y,global}$ is the force acting on and
A is the area of the load-carrying plate

In addition, wherever relevant, the stresses $\sigma_\parallel$ in the load-carrying plate shall also be determined. For detail 3 from Table 8.5 of EN 1993-1-9, subject to both normal and shear stresses, this translates into the relationships (3.26) to (3.28). The corresponding possible fatigue crack cases are represented in Figure 3.15. In the case of fillet welds, fatigue cracking usually occurs from the root (case A) and thus only this case must be verified. Case B represents the nominal stress in the gross section of the loaded plate.

Case A:
(from root)

$$\sigma_w = \sqrt{\left(\frac{F_z}{2a_{eff} \cdot \ell}\right)^2 + \tau_\perp^2} \qquad (3.26)$$

$$\tau_w = \tau_\parallel = \frac{F_y}{2a_{eff} \cdot \ell} \qquad (3.27)$$

Case B:
(from toe)

$$\sigma = \frac{F_z}{t \cdot \ell} \qquad (3.28)$$

where

F_y tangential force on the loaded attachment,
F_z axial force on the loaded attachment,
ℓ length of the attachment,
a_{eff} effective weld throat for partial penetration welds, or weld throat a,

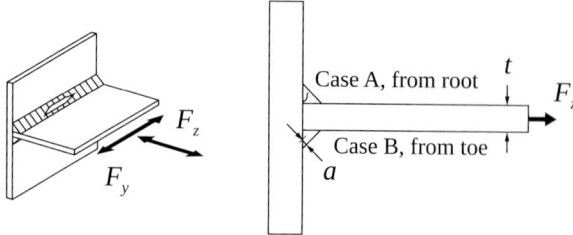

Figure 3.15 – Partial penetration T-butt joints or fillet welded joint (double-sided)

In the case of Table 8.5, detail 9, this translates into the following relationship (3.29), see also Figure 3.16:

3.3 CALCULATION OF STRESSES

$$\tau_w = \tau_{\parallel} = \frac{F_z}{2 \cdot 2a \cdot \ell} \qquad (3.29)$$

where
    F_z total force on the connected plates composing the double-sided lap joint.

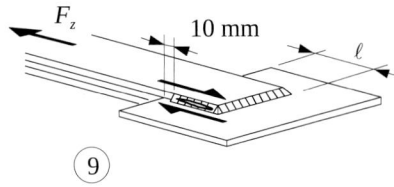

Figure 3.16 – Fillet welded double-sided lap joint

3.3.5 Nominal stresses in steel and concrete composite bridges

The first step for determining the nominal stresses in a structural detail is to perform the elastic global cracked analysis of the composite steel and concrete bridge according to EN 1994-2, and to calculate the internal forces and moments for the basic SLS combination of the non-cyclic loads which is defined in EN 1992-1-1, § 6.8.3:

$$G_{k,sup} \left(\text{or } G_{k,inf} \right) + \left(1 \text{ or } 0 \right) S + 0.6 T_k \qquad (3.30)$$

where
    G_k characteristic nominal value of the permanent actions effects,
    S characteristic value of the effects of the concrete shrinkage,
    T_k characteristic value of the effects of the thermal gradient.

The non-structural bridge equipments (safety barriers, asphalt layer,...) have to be calculated by integrating an uncertainty on the characteristic value of the corresponding action effects. The corollary is that two values of the internal forces and moments, a minimum and a maximum one, have to be considered in every cross section of the composite bridge. Each bound of this basic envelope should be considered independently for adding the effects of the fatigue load model (usually FLM3 from EN 1991-2) in the combination of actions.

For the second step of the calculation of the nominal stresses, the bridge design specifications should settle the number and the location of the slow traffic lanes on the bridge deck. These assumptions are then used for calculating the transversal distribution coefficient for each lane. The FLM3 crossing the bridge in the slow lane induces a variation of the internal forces and moments in the bridge, which should be added to the maximum (resp. minimum) bound of the envelope for the basic SLS combination of non-cyclic actions. Two different envelopes, named case 1 and 2 in the following, are then defined:

- case 1:

$$\min\left[G_{k,sup}\left(\text{or } G_{k,inf}\right)+\left(1 \text{ or } 0\right)S+0.6T_k\right]+FLM3 \quad (3.31)$$

- case 2:

$$\max\left[G_{k,sup}\left(\text{or } G_{k,inf}\right)+\left(1 \text{ or } 0\right)S+0.6T_k\right]+FLM3 \quad (3.32)$$

The calculation of the nominal stresses should be performed for each bound of these two cases (it means that 4 different values of the internal forces and moments have to be considered, finally leading to 2 values for the stress range). The stress calculation should take the construction sequences into account and if one of these bending moments induces a tension in the concrete slab, the corresponding stress value should be calculated with the cracked properties of the cross section resistance.

In this Manual and according to the Eurocode notations, for both previous cases, the bounds of the envelope for the bending moment are noted by pairs $M_{Ed,max,f}$ and $M_{Ed,min,f}$ respectively.

See section 3.7.6 for more details about the stress range calculations in a composite bridge, and its application to a numerical example.

3.3.6 Nominal stresses in tubular structures (frames and trusses)

This section is dedicated to a special case of determining nominal stresses in tubular trusses, frames or lattice girders with directly welded tube joints. In the case of trusses, because of node stiffnesses, secondary bending moments exist in lattice girder joints (e.g. K- and N-joints). For static design, these moments are not important if the critical members or joints have sufficient rotation capacity. A structural truss model with pinned nodes can thus be used. However, for fatigue design, the peak stress (and resulting stress range) is the governing parameter and secondary bending moments influence those significantly. As a consequence, secondary bending moments have to be considered in fatigue design. Secondary bending moments are caused by various influences, such as:

- overall bending stiffness of the joint,
- local joint flexibility (i.e. stiffness distribution in the joint along the intersection perimeter),
- eccentricities between the members at the node.

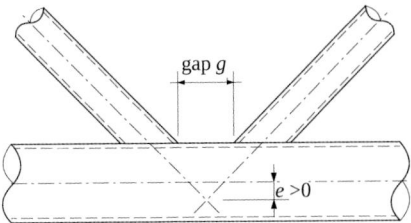

Figure 3.17 – Positive eccentricity in a K-joint made of circular hollow sections (CIDECT, 2001)

Figure 3.17 shows, as an example, a K-joint where the centre lines of the braces do meet below the centre line of the chord. This situation is called positive eccentricity and results in a secondary bending moment. Note that a negative eccentricity with overlapping braces is also possible.

Regarding fatigue analyses, secondary bending always has to be considered – at least for the chord as most often fatigue cracks develop in it. This can be done using one of the four following analysis:

1) A simplified truss model analysis (only pin joints) is made. This modelling is only valid for joints without or with small eccentricities (typically less than 2 % of diameter). Then, the nominal stress ranges obtained for axial loading are to be multiplied by correction factors, denominated k_1. This results in what is actually a modified nominal stress (see sub-chapter 3.4). These factors account for the secondary bending moment effects from joint stiffness since they are not included in the analysis. In EN 1993-1-9, Table 4.1 (for circular hollow section joints) and Table 4.2 (for rectangular hollow section joints) contain the k_1–factor values, given in function of the joint type (N-joints, T-joints, K-joints, or KT-joints). The k_1–factor values given in the tables are upper bound values based on measurements in actual girders and tests as well as finite element calculations. With modern

3. DETERMINATION OF STRESSES AND STRESS RANGES

computers and powerful finite element modelling software, this method tends to be replaced by the other methods presented below. Thus, the use of this method should be limited to the lattice girder node joints given in EN 1993-1-9, Table 8.7 (modified stress method, see sub-chapter 3.4, and corresponding fatigue check).

2) A frame analysis for triangulated trusses or lattice girders is used as before. In this case, it is modelled by considering a continuous chord with brace members pin connected to it at distances of $+e$ or $-e$ from it (e being the distance from the chord centreline to the intersection of the brace member centrelines) (CIDECT, 2001)(IIW, 2000). Excentricity values can take values up to 25 % of the diameter according to static strength design rules (IIW, 2009b). The links to the pins are treated as being extremely stiff as indicated in Figure 3.18. The advantage of this model is that a sensible distribution of bending moments is automatically generated throughout the truss, for cases in which bending moments need to be taken into account in the design of the chords (Wardenier, 2011). This results in axial forces in the braces, and both axial forces and bending moments in the chord. It accounts for excentricity effects, but not for joint stiffness effects. Thus, as in method 1, the stress ranges in chord and braces caused by the axial loading in the braces, and those only (not the bending stresses) have to be multiplied by the factors given by correction factors k_1 given in Table 4.1 and Table 4.2 of EN 1993-1-9. Modified nominal stresses are obtained, which usually will be multiplied by SCF (Stress Concentration Factors, see sub-chapter 3.4) in order to use a geometric stress method, see sub-chapter 3.5, and corresponding fatigue check. This method has been validated by comparison between measurements in actual girders and finite element calculations (Wardenier, 2011). The authors think this method has a drawback since the physical meaning is lost once axial stresses are increased using correction factors k_1 and thus "simulate" maximum secondary bending stress effects, also in the chord which indeed has already bending stresses from analysis.

3) A rigid frame analysis is used. That is, a continuous chord with braced members rigidly connected to it at distances of $+e$ or $-e$ from it, the links to this node being again treated as being extremely stiff as indicated in Figure 3.19. The bending moments in the braces are to be

taken at a distance $D/2$ from the chord member, i.e. at the points corresponding to the chord tube surface. The stresses deduced from the axial forces and bending moments do not need to be multiplied by correction factors. This method has also been validated by comparison between measurements in actual girders and finite element calculations (Romeijn et al, 1997) (Walbridge, 2005). It is especially suited for bridges with K-joints or KK-joints where the members tend to be stocky (low ratios $\gamma = D/2T$). Modified nominal stresses are obtained, which usually will be multiplied by SCF factors in order to use a geometric stress method fatigue check. It should however be mentioned that according to Wardenier (2011), this type of analysis may exaggerate brace member moments, while the axial force distribution will still be similar to that for a pin jointed analysis.

4) Finally, joints can be modelled as three dimensional substructures using shell or solid finite element modelling. When, among others, stiffnesses and boundary conditions are correctly modelled, secondary bending moments can be accurately taken into account. However, such modelling is only appropriate for experienced analysts. In this case, the structural stress at the hot spot can be found directly by stress extrapolation (see for example Niemi et al (2006) for extrapolation methods). One can then use a geometric stress method fatigue check.

The third method is seen by the authors as the most efficient and economical. For two- or three-dimensional Vierendeel type girders, all the same, rigid frame analyses are recommended.

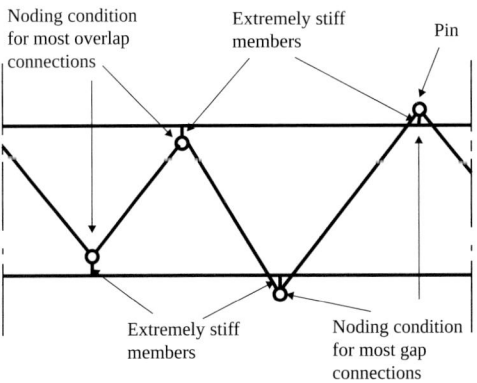

Figure 3.18 – Plane frame analysis according to CIDECT (CIDECT, 2001)

3. DETERMINATION OF STRESSES AND STRESS RANGES

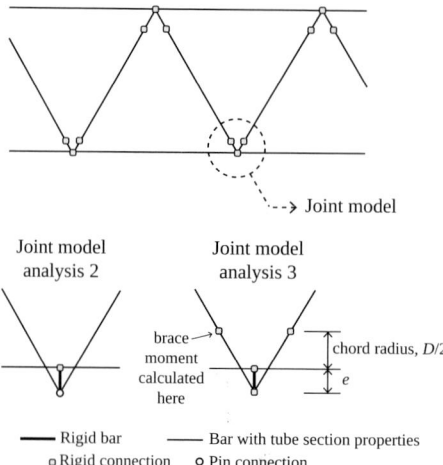

Figure 3.19 – Plane frame analyses and joint modelling assumptions (Walbridge 2005)

3.4 MODIFIED NOMINAL STRESSES AND CONCENTRATION FACTORS

3.4.1 Generalities

This method of determining the stress distribution corresponds to a refinement from the nominal stress methods presented before. With this method, the designer can take into account the effect of geometric stress concentrations which are not a characteristic of the detail category itself, such as for example:

- holes (Figure 3.20) and cut-outs,
- re-entrant corners,
- eccentricities or misalignments not accounted for previously.

The resulting geometric stress concentration relevant for fatigue design should be determined either by special structural analysis or, where appropriate, by the use of predefined fatigue stress concentration factors.

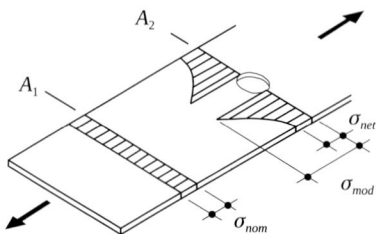

Figure 3.20 – Geometric stress concentration in the vicinity of a hole

3.4 MODIFIED NOMINAL STRESSES AND CONCENTRATION FACTORS

The following relationship is used to compute the modified nominal stress:

$$\sigma_{mod} = k_f \cdot \sigma_{nom} \qquad (3.33)$$

where k_f is the geometric stress concentration factor (SCF). Note that in EN 1993-1-9, the geometric stress concentration factor is directly included in the relationship for the stress range (see section 3.7.7).

Handbooks with geometric stress concentration factors exist but they are made for mechanical engineering applications (shafts, discs, etc.), usually referred as k_t in the specialized literature, and may not contain the cases needed for structural applications. The following handbooks and references may be used:

- British standard (BS 7608, 1993) for openings and re-entrant corners, see Figure 3.21 and Figure 3.22;
- British standard (BS 7910, 1999) for eccentricities and misalignments.
- Det Norske Veritas (DNV, 2010) for manholes and stiffened openings;

Note that the geometric stress concentration may refer to either the nominal stress in the gross section or the nominal stress in the net section (given as σ_{net} in Figure 3.20).

Eccentricities and misalignment have to be accounted for through the use of an additional geometric stress concentration factor k_f. This can be done either by increasing the stress range (action effects side of verification) using expression (3.33), which would be logical, or by reducing the fatigue detail category (strength side). The second method is unfortunately used in the Eurocodes, thus mixing factors resulting from action effects with fatigue strength, using expression (3.34) given below.

$$\Delta\sigma_{C,red} = \frac{1}{k_f} \cdot \Delta\sigma_C \qquad (3.34)$$

where
- $\Delta\sigma_{C,red}$ Reduced fatigue detail category
- $\Delta\sigma_C$ Original fatigue detail category

The reason given above explains why, even if the factor appears on the resistance side, this aspect is mentioned in this section. It is further

3. DETERMINATION OF STRESSES AND STRESS RANGES

explained in section 3.7.7. In fact, in other international codes, such as IIW recommendations (IIW, 2009), British standard (BS7608, 1993) or in the domain of offshore structures (DNV, 2010), the influence of possible eccentricities is always accounted for using expression (3.33), that is by an increase on the action effects side.

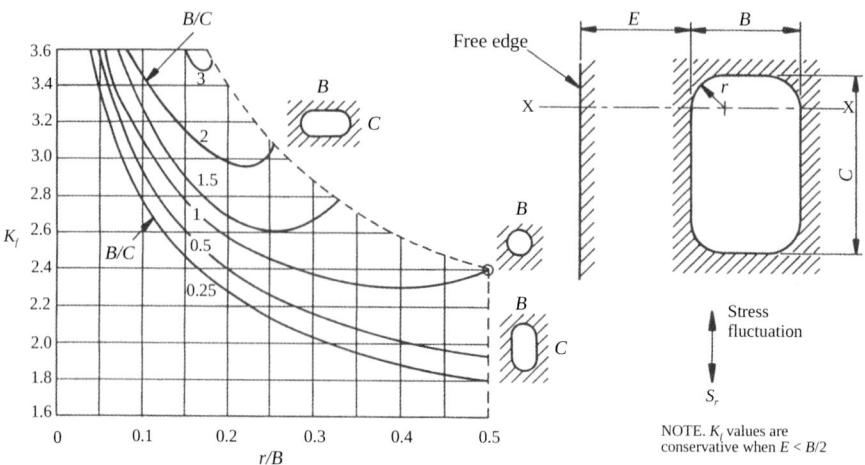

Figure 3.21 – Geometric stress concentration factors at holes and unreinforced apertures (based on net stress at X-X) adapted from BS 7608 (BS 7608, 1993)

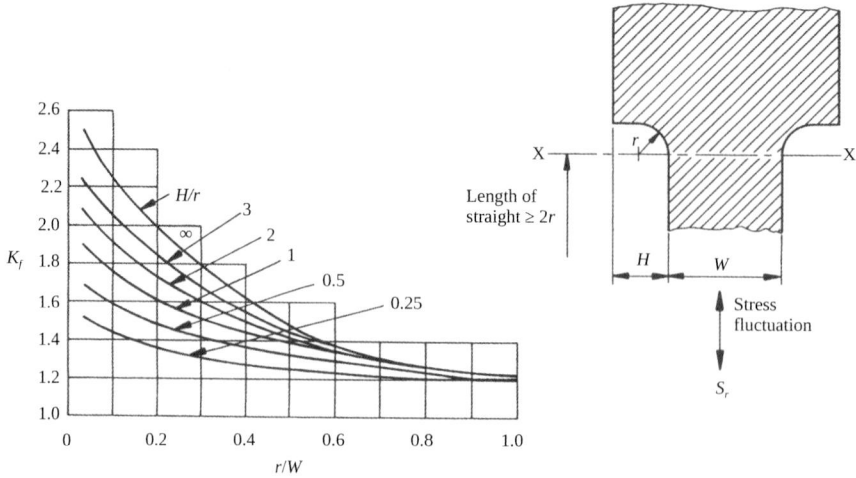

Figure 3.22 – Geometric stress concentration factors at re-entrant corners (based on net stress at X-X), adapted from (BS 7608, 1993)

3.4 MODIFIED NOMINAL STRESSES AND CONCENTRATION FACTORS

3.4.2 Misalignments

For typical fabrication tolerances (e.g. eccentricities or angular misalignment), the following paragraphs give a summary of the geometric stress concentration factor formulas that can be used.

An eccentricity in welded connections loaded axially results in additional stresses in the form of secondary bending. To summarise, three different cases are identified (see Figure 3.23):

  a) axial misalignment between the centroidal axes from plates of identical nominal thicknesses (e.g. in butt welds),
  b) axial misalignment between the centroidal axes from plates of different thicknesses (e.g. in bridge flanges butt welds),
  c) axial misalignment in cruciform joints (e.g. in orthotropic decks stringer–crossbeam–stringer connections).

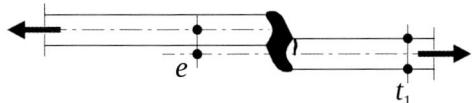

a) Axial misalignment between identical plates

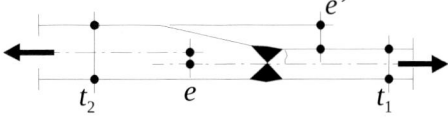

b) Axial misalignment between plates of different thicknesses

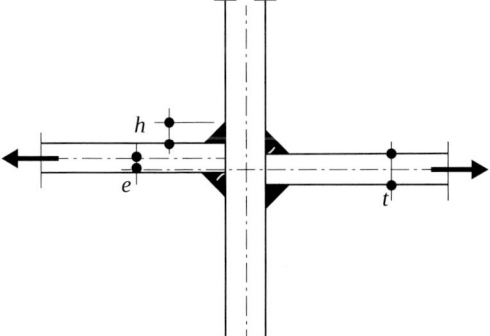

c) Axial misalignment in cruciform joints

Figure 3.23 – Possible cases of axial misalignment between plates

3. Determination of Stresses and Stress Ranges

For the case a), the additional bending stresses can be computed using the following geometric stress concentration factor expression, taken from the British Standard (BS 7910, 1999):

$$k_f = 1 + \frac{3e}{t_1} \quad (3.35)$$

where

t_1 plate thickness,
e misalignment or eccentricity, see Figure 3.23a).

In DNV (2010), the misalignment between plates e is replaced by the expression $(e-e_0)$, where e_0 represents a misalignment equal to $0.1t$, which corresponds to the misalignment value already accounted for in the DNV detail classification for transverse butt welds. The authors believe that a similar approach could be used for EN 1993-1-9 details.

For the case b), the additional geometric stress concentration factor can be expressed as follows:

$$k_f = 1 + \frac{6e}{t_1} \cdot \frac{t_1^{1.5}}{t_1^{1.5} + t_2^{1.5}} \text{ with } t_1 \leq t_2 \quad (3.36)$$

where

t_1 thickness of the thinner plate,
t_2 thickness of the thicker plate,
e misalignment or eccentricity, see Figure 3.23b).

Since in practice the value e of the misalignment between plates cannot be directly measured, the following equation can be used instead:

$$e = e' - \frac{1}{2} \cdot (t_2 - t_1) \quad (3.37)$$

where

e' plates misalignment in function of the difference between flat surfaces, see Figure 3.23b)

As for case a), in DNV (2010) the misalignment between plates e is replaced by the expression $(e-e_0)$, where e_0 represents a misalignment equal to $0.1t_1$ this time, t_1 being the thickness of the thinner plate.

For bridge girders, the additional stress concentration resulting from misalignments in transverse flange butt welds may in general be neglected as

3.4 MODIFIED NOMINAL STRESSES AND CONCENTRATION FACTORS

t_2/t_1 is less than or equal to 2.0. The reason is that it is a case of plates that are supported (by the web), which significantly reduces the misalignment secondary moment. The use of the geometric stress concentration factors presented herein would lead to a very conservative design. For other cases, a recent study has been completed (Lechner and Taras, 2009).

For the case c), the additional geometric stress concentration factor resulting from a misalignment in cruciform joints with fillet welds can be expressed as follows:

$$k_f = 1 + \frac{e}{t+h} \qquad (3.38)$$

where
- t thickness of the attached plates,
- h fillet weld leg size, see c), usually taken as $h = \sqrt{2} \cdot a$,
- e misalignment or eccentricity.

Equation (3.36) can be applied to account for wall eccentricities in butt welds between tubes with thickness transition on the outside (Maddox, 1997; McDonalds and Maddox, 2003). However, an improvement of the geometric stress concentration factor expression for tubes has been proposed by Lotsberg (2009) and is given in equation (3.39).

$$k_f = 1 + \frac{6(e + e'')}{t_1} \cdot \frac{1}{1 + (t_2/t_1)^\beta} \exp^{-\alpha} \qquad (3.39)$$

with

$$\alpha = \frac{1.82 \cdot L}{\sqrt{D \cdot t_1}} \cdot \frac{1}{1 + (t_2/t_1)^\beta}$$

and

$$\beta = 1.5 - \frac{1.0}{\log(D/t_1)} + \frac{3.0}{(\log(D/t_1))^2}$$

where
- t_1 thickness of the thinner tube,
- t_2 thickness of the thicker tube,
- e misalignment or eccentricity of tube wall, see Figure 3.23 b),
- e'' misalignment in tube axis, taken positive if in same direction as e,
- D thinner tube outside diameter
- L length of transition in thickness.

3. DETERMINATION OF STRESSES AND STRESS RANGES

The following limits have to be respected:

- $t_2/t_1 \leq 2$;
- $20 \leq D/t_1 \leq 1000$ and
- $L/(t_2 - t_1) \geq 4$.

For butt welds between tubes with thickness transition on the inside, equation (3.40) may be used but the modified stress range and crack location are located on the inside face (at the weld root).

Finally, one can also have an angular misalignment between plates as represented in Figure 3.24. This latter case can be accounted for using the following relationship (Hobbacher, 2003):

$$k_f = 1 + \beta_r \cdot \alpha \frac{L}{2t} \qquad (3.40)$$

where

- L unsupported length,
- α angular misalignment in degrees,
- β_r factor to account for end conditions,
- t plate thickness.

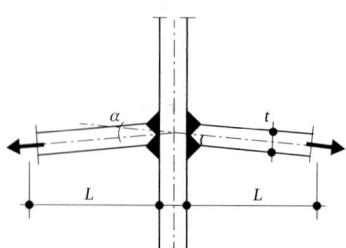

Figure 3.24 – Case of angular misalignment

Example 3.6: Application to chimney, fatigue verification of manhole detail (worked example 2)

In this example, the fatigue verification of the manhole near the bottom of the chimney will be performed. The geometry and dimensions are given in section 1.4.3. The manhole corresponds to a cut-out in the shell; its dimensions and shape are given in sub-section 1.4.3.5 and Figure 1.21. The bending moment range at the bottom level of the manhole was calculated in Example 3.1 as: $\Delta M_3 = 648.6$ kNm. Two cases are presented: an unreinforced manhole and a reinforced one (see Figure 3.25, case taken from DNV RP 203 (DNV, 2010), annex C2, reinforcement type D). The wind loads are computed in Example 3.1.

Figure 3.25 – Reinforcement of the manhole edge and positions of computed modified stresses

Loading

Unreinforced case (accounting for reduction in section modulus due to manhole):

Using the section properties at the level of the manhole and assuming the wind acts in the most unfavorable direction, the stress range along the edge of the manhole, conservatively taken at the bottom (without geometric stress concentration factor) is:

$$\Delta\sigma_{E,mh,net} = \frac{\Delta M_3}{W_{y,mh}} = \frac{648.6 \cdot 10^6}{20000 \cdot 10^3} = 32.4 \text{ N/mm}^2$$

Note: in this case, a net section stress range is needed.

Reinforced case:

The stress range, computed in again at bottom of the manhole but with full section since it is reinforced (without geometric stress concentration factor), is:

$$\Delta\sigma_{E,mh} = \frac{\Delta M_3}{W_y} = \frac{648.6 \cdot 10^6}{24493 \cdot 10^3} = 26.5 \text{ N/mm}^2$$

Regarding the influence of the manhole on fatigue, EN 1993-3-2, section 9.1, says the following: where the geometrical stress method is used, such as at openings or by a particular shape of connection, geometric stress concentration factors may be used according to EN 1993-1-6. None of the Eurocodes does however give geometric stress concentration factors, thus literature has to be used.

3. Determination of Stresses and Stress Ranges

Modified stress range computation:

Unreinforced case:

The geometric stress concentration factors are obtained from BS 7608, also given in Figure 3.21.

Geometrical parameters for the manhole:

$$\frac{W_{mh}}{h_{mh}} = \frac{B}{C} = \frac{600}{1200} = 0.5$$

$$\frac{r_{mh}}{W_{mh}} = \frac{B}{C} = \frac{300}{600} = 0.5$$

$$h_{mh} = 1200 \text{ mm}$$

$$W_{mh} = 600 \text{ mm}$$

$$r_{mh} = 300 \text{ mm}$$

From Figure 3.21, $k_f = 1.95$

The modified stress range along the edge of the manhole is:

$$\Delta\sigma_{E,mh,mod} = k_f \cdot \Delta\sigma_{E,mh,net} = 1.95 \cdot 32.4 = 63.2 \text{ N/mm}^2$$

Reinforced case:

The geometric stress concentration factors are obtained from DNV RP-C203 (DNV, 2010), Annex C2, reinforcement type D.

Position 1)

$$k_{f1} = 2.6$$

$$\Delta\sigma_{E,mh1,mod} = k_{f1} \cdot \Delta\sigma_{E,mh} = 2.6 \cdot 26.5 = 68.9 \text{ N/mm}^2$$

Position 2)

$$k_{f2} = 1.65$$

$$\Delta\sigma_{E,mh2,mod} = k_{f2} \cdot \Delta\sigma_{E,mh} = 1.65 \cdot 26.5 = 43.7 \text{ N/mm}^2$$

Position 3) bottom end of stiffener, no geometrical stress concentration

$$\Delta\sigma_E = \Delta\sigma_{E,mh} = 26.5 \text{ N/mm}^2$$

As explained in Example 3.5, due to the large number of load cycles ($> 10^8$), the only possible verification to satisfy fatigue design is to do it with the fatigue limit (see section 5.4.2).

3.5 GEOMETRIC STRESSES (STRUCTURAL STRESS AT THE HOT SPOT)

3.5.1 Introduction

This method of determining stresses corresponds to the more refined way of determining stress distribution. In this method, the so-called *geometric stress* or *structural stress at the hot spot*, σ_{hs}, includes all stress raising effects of a structural detail apart from the weld itself. The structural stress at the hot spot is the stress at the plate surface and the weld toe, where the fatigue crack is expected and where joint failure will start. It includes the effects of joint geometry and the type of load (global effects), but excludes local effects due to the weld shape, radius of the weld toe (notch effects), etc., see Figure 3.26. It corresponds to the membrane stress plus shell bending stress at the hot spot.

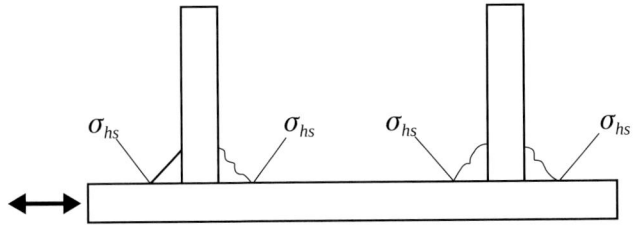

Figure 3.26 – View of welded transverse attachments with different weld shapes, all having the same structural stress at the hot spot, σ_{hs}

The field of application of the geometric stress approach is welded joints for which (Niemi *et al*, 2006):

- the fluctuating principal stress acts predominantly transverse to the weld toe (or the ends of a discontinuous longitudinal weld),

3. DETERMINATION OF STRESSES AND STRESS RANGES

- the potential fatigue crack will initiate at the weld toe or end (in other words not from the weld root).

Since the structural stress at the hot spot is impossible to measure or calculate at the weld toe, standard procedures have been developed for determining it. They all involve the extrapolation of stresses from a region adjacent to the weld toe to the weld toe itself. This concept is illustrated in Figure 3.27.

The structural stress at the hot spot can be seen as the highest structural stress in the region of a weld, or a potential crack location. Since it accounts for the stress concentrations resulting from the detailing, it allows for regrouping different structural geometries into a single case for fatigue design. Indeed, in this computation method, macro- and microscopic effects on fatigue are separated. Macroscopic effects, i.e. stress concentrations, are included in the calculated geometric stress range and the microscopic effects (e.g. weld shape, weld type, flaws, etc.) are built into a reduced set of empirical hot spot S-N curves, which are given in EN 1993-1-9, annex B.

The boundaries of the extrapolation region, within which the stresses used in the determination should be measured or calculated, are defined for all cases in IIW (2000) and for tubular joints in CIDECT (2001). The extrapolation of the stress to the weld toe by using the stresses values, either measured or computed, in a zone adjacent to the weld is shown in Figure 3.27. In case of FEM modeling, note that different rules have to be followed if the model is made out of shell or solid elements.

The modified nominal stress and the geometric stress methods are a lot alike. However, in general, the purpose of the two methods is different:

- The modified stress approach is intended for cases where a detail exists in the detail category tables but with an additional SCF. This SCF is linked with the geometry of the element, not with the geometry of the weld.
- The geometric stress method is intended for welded details not listed in the detail category tables or for details with complicated stress fields in the vicinity of the weld detail where the crack starts.

Both methods use SCF's, which represent the ratio between the stress value that is governing fatigue cracking and the nominal stress value (away from the crack). In special cases, the product of SCF's can be made to obtain the geometric stress value. These concepts are developed in the next sections.

3.5 GEOMETRIC STRESSES

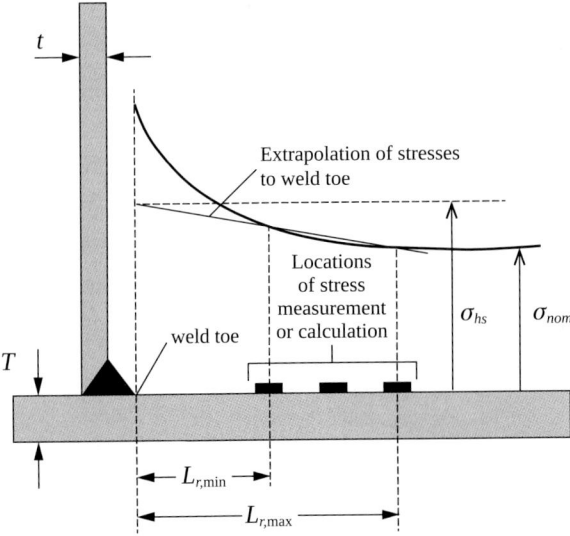

Figure 3.27 – Illustration of concept of extrapolation of surface stresses to weld toe (Schumacher, 2003)

3.5.2 Determination using FEM modelling

Structural stress at hot spots are stresses which are determined in the structure with the assumptions of the classical engineering theories (elastic behaviour, linearized stress distribution over plate thickness and others). The notch effect is more or less suppressed, but the local stress increase originating from the macro shape is included. These stresses are either measured or calculated because of the singular point in numerical models and the impossibility to place gages at this point in experimental specimens. The structural stress at the hot spot can be obtained numerically by extrapolation, using finite element method models (FEM), or boundary element method models (BEM). Calculation is based on engineering formulae or finite element analysis, normally using rather coarse meshes. The questions to be answered are: where are the failure-critical points ('hot spots') in the structure? and which geometric stress can best describe the local failure condition? The conventional way of determining the structural geometric stress at the hot spot is the extrapolation of stresses on the plate surface to the weld toe, as already shown in Figure 3.27. The assumption on linearized stress distribution over plate thickness results from the basic idea behind the geometric stress where the non-linear stress peak due to the weld

3. DETERMINATION OF STRESSES AND STRESS RANGES

is excluded and allows for the determination of the geometric stress with both shell element models or solid element models. Specific rules for each of the FEM model type are necessary and can be found in different recommendations, in function of the domain of application.

Two types of structural stress at the hot spot can be distinguished according to their location on the plate and their orientation to the weld toe (Niemi *et al*, 2006):

- Hot spot type *a* – structural stress transverse to weld toe on plate surface,
- Hot spot type *b* – structural stress transverse to weld toe at plate edge.

The determination of each of these two structural stresses types involves the extrapolation using stress values at several points in the vicinity of the weld toe. Figure 3.28 shows the distances at which the extrapolation should be done depending on the mesh size. But it is not the purpose of this manual to give precise guidelines for determining the geometric stress; those guidelines on finite element modelling of the structure, of the welds, and evaluation of the stresses are available in Niemi *et al* (2006) and DNV (2010). An alternative procedure in the case of models with solid elements is the internal linearization of the stresses over the plate thickness at the weld toe.

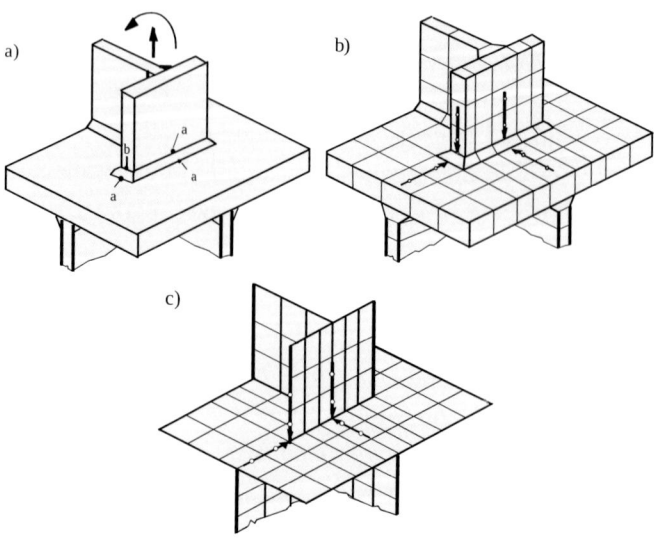

Figure 3.28 – Types *a* and *b* of structural stress at the hot spot obtained from FEA by extrapolation on surface using solid elements (b) and shell elements (c) (extracted from (Niemi *et al*, 2006))

3.5 GEOMETRIC STRESSES

A FEM model is usually an idealisation of the actual geometry of the structure, meaning that in reality the structure will be different due to fabrication tolerances and welding distortions. Thus, if the FEM analysis is carried out on an ideal geometry, the nominal stress found can be multiplied by a magnification factor to first get an estimation of the modified nominal stress, σ_{mod}, according to the following formula:

$$\sigma_{mod} = k_f \cdot \sigma_{nom,ax} + \sigma_{nom,b} \qquad (3.41)$$

where
- k_f geometric stress concentration factor, for example from misalignment,
- $\sigma_{nom,ax}$ axial or membrane part of the nominal stress,
- $\sigma_{nom,b}$ bending part of the nominal stress.

The resulting modified stress is then used, instead of the nominal stress, in formula (3.42) to get the geometric stress value, σ_{hs}.

Note that there exist other specific methods. The ASME's boiler and pressure vessel code adopted in 2007 an alternative method for determining the structural stress. It is a proprietary, mesh-insensitive, structural stress method developed by Dong (2001). The DNV (2010) also recommends the use of the effective notch stress method, with the so-called ficticious notch radius, in particular for the FEM modelling of fatigue details with complex geometries or in case of a fatigue crack developing from the weld root.

3.5.3 Determination using formulas

In general, the structural stress at the hot spot is determined using the same concept of stress concentration factor already presented in the section on modified nominal stress:

$$\sigma_{hs} = k_{f,hs} \cdot \sigma_{nom} \qquad (3.42)$$

where
- σ_{hs} structural stress at the hot spot,
- $k_{f,hs}$ geometric stress concentration factor, to simplify also noted as k_f,
- σ_{nom} nominal stress value remote from the detail.

3. Determination of Stresses and Stress Ranges

Concentration factor values for the structural stress at the hot spot can be found in the literature for a large number of different details. These factors are obtained analytically or from parametric studies. Note that in many cases, the geometric stress concentration factors for the axial and bending loading components are given separately.

One such set of geometric stress concentration factors formulae exists for the design of welded tubular joints. Its base is the most extensive study to date for both uni-planar and multi-planar joints, based on both experimental results and FEM analyses (using solid elements for the entire joint model), carried out by Koning et al (1992), Romeijn (1994) and Romeijn et al (1997). A database of geometric stress concentration factor results from this work was analysed and used to establish the most up-to-date design specifications for fatigue in tubular joints. For K-joints specifically, the hot spot stress concentration factor equations cover gap joints without eccentricity within specified validity ranges (IIW, 2000).

In the case of complex geometries and loadings, a typical case being tubular joints, combination of different so-called basic load cases is necessary as follows:

$$\sigma_{hs} = k_{f1} \cdot \sigma_{nom,1} + k_{f2} \cdot \sigma_{nom,2} + k_{f3} \cdot \sigma_{nom,3} + \ldots \quad (3.43)$$

where

σ_{hs} structural stress at the hot spot under a given load combination,

k_{fi} geometric stress concentration factor for basic load case i,

$\sigma_{nom,i}$ nominal stress value remote from the detail for basic load case i (can be an axial stress or a bending stress according to the load case definition).

For tubular joints, the CIDECT Fatigue Design Guide (CIDECT, 2001) or the IIW (2000) recommendations give formulae and graphs for different types of joints. Each formula is valid for a specific basic load case and is a function of the tubular joint geometric parameters. For example, in the case of the balanced axial loading, the formula reads as follows:

$$k_{f,j,ax} = SCF_{j,ax} = \left(\frac{\gamma}{12}\right)^{0.4} \cdot \left(\frac{\tau}{0.5}\right)^{1.1} \cdot SCF_{0,j,ax} \quad (3.44)$$

where
- j chord or brace,
- $k_{f,j,ax}$ geometric stress concentration factor (in the chord or in the brace) due to the basic load case balanced axial loading (ax),
- γ geometric parameter, ratio chord radius to chord thickness,
- τ geometric parameter, ratio brace to chord thicknesses,
- $SCF_{0,ax}$ reference value for the stress concentration factor (in the chord or in the brace) due to balanced axial loading.

Another example, for the basic load case of chord loading (axial and in-plane bending), the stress concentration formula reads (CIDECT, 2001):

$$k_{f,j,ch} = SCF_{j,ch} = 1.2 \cdot \left(\frac{\gamma}{0.5}\right)^{0.3} \cdot (\sin\theta)^{-0.9} \qquad (3.45)$$

where
- j chord or brace,
- $k_{f,j,ch}$ stress concentration factor (in the chord or in the brace) due to the basic load case chord loading (ch),
- γ geometric parameter, ratio chord radius to chord thickness,
- θ angle between the chord and the braces.

3.6 STRESSES IN ORTHOTROPIC DECKS

Ideally, in orthotropic decks, one should compute the structural stress at the hot spot values at each detail. But the problem is very complex; often only nominal stresses are computed and only detail categories with respect to nominal stresses are given. Even for nominal stresses, determining them in orthotropic decks near details is very difficult since the load carrying system is composed of four different types of members interacting together, namely: the deck plate, the longitudinal open or closed stiffeners (referred to as troughs or also stringers), the crossbeams and the main girders. A short description of a typical orthotropic deck system and its behaviour is needed in order to understand how one shall compute correctly, without a detailed FEM model, the stresses in such a system. A typical orthotropic steel deck system is shown in Figure 3.29.

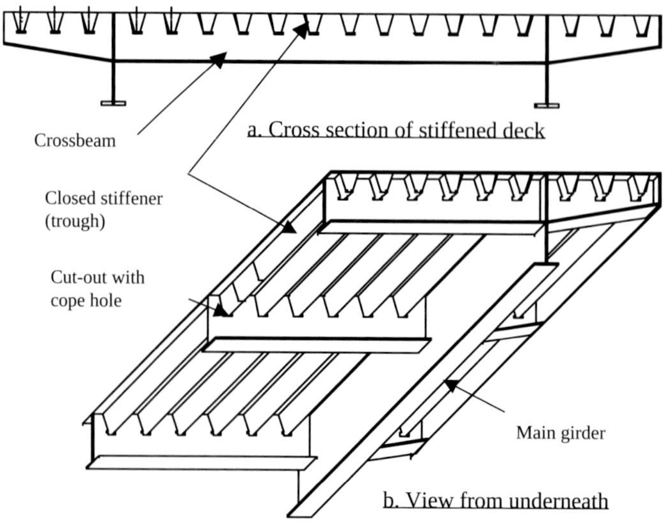

3. Determination of Stresses and Stress Ranges

Figure 3.29 – Typical orthotropic steel deck with crossbeams and main girder (Leendertz, 2008)

The behaviour of an orthotropic steel deck system can be summarized as follows (Leendertz, 2008):

- Vertical traffic loads are applied to the wearing course of the steel deck and hence transferred to the steel deck plate, which is supported longitudinally by the stiffeners (flat stiffeners or, as shown in Figure 3.29, trapezoidal stiffeners). The deck bends in the bridge transverse direction. Shear and bending stresses are thus generated in the transverse direction of the deck plate.
- In the longitudinal direction, the stiffeners act together with a part of the deck plate and transfer the applied traffic loads to the crossbeams. Shear forces and bending moments are present in the stiffeners.
- The stiffener supports (stiffener to crossbeam connections) transfer the applied loads to the crossbeams. Due to the deflection of the stiffeners between the crossbeams, the supports are subject to a rotation. This results in an out-of-plane displacement of the crossbeam web, the so-called *out-of-plane crossbeam behaviour*.
- In the transverse direction, the crossbeams act together with a part of the deck plate acting as the upper flange and transfer the stiffener supports reaction forces to the main beams. The load transfer in the crossbeam generates shear forces and bending moments under the

in-plane crossbeam behaviour. It is important to mention that since the crossbeam web contains cut-outs as stiffeners are usually continuous through them, significant in-plane shear deformations (and stresses) may occur in the crossbeam. Furthermore, as mentioned before, the crossbeams will be deformed by the applied rotations of the stiffeners caused by bending under traffic loads, which causes an out-of-plane displacement and thus local bending and torsion in the crossbeam web.

The behaviour, load transfer and stress distribution in an orthotropic steel deck system is strongly affected by its type, geometric proportions and different details (open or closed stiffeners, stiffener and cut-outs shape, stiffener to crossbeam connection, etc.). In modern bridge design and rehabilitation, the trend seen by the authors is towards two different so-called "optimum deck fatigue designs". They are based on historical development, country construction habits and experimental as well as numerical validations. They can be described as follows:

- Short spacings between crossbeams (between 2 and 3 m) thus giving minimum crossbeam and stiffener height. The stiffener width is usually equal to its height and spacing ranges between 300 and 600 mm. This design is often used, among other countries, in the United States and Japan. In order to have good fatigue behaviour, this design requires cut-outs to be as small as possible and to add reinforcing plates in the stiffeners to improve the crossbeam in-plane behaviour and reduce in-plane shear deformations. The applied rotations of the stiffeners are small because of their short spans and their small height; thus out-of-plane crossbeam deformations are limited.
- Large spacings between crossbeams, typically ranging from 3 to 5 m, thus giving a minimum number of crossbeams. The crossbeams are deeper, more stiffeners are needed than in the other optimal design and their height is greater. The stiffeners are higher than they are wide and spacing ranges between 600 and 900 mm. This design is usually used in Europe (especially in Germany and the Netherlands) and recommendations for the dimensions and detailing are given in Annex C of EN 1993-2. In order to have good fatigue behaviour, this design requires large cut-outs as applied rotations of the stiffeners are large because of the long stiffener spans and their heights. The

in-plane behaviour of the crossbeam is satisfactory because of its height, resulting in limited in-plane shear deformations.

Orthotropic decks contain different fatigue details and a distinction can be made between cracks that are caused in a load-carrying member or in a connection for load transfer, and cracks that are generated by imposed deformations. Also, depending upon the details, either a nominal or a hot spot stress approach is recommended in the code. In EN 1993-2, there is some specific information on orthotropic decks: Annex C contains information on behaviour and 9.4.2 contains information on analysis for fatigue and stress determination. Further information about fatigue strength of orthotropic deck details is given in section 4.2.8 and in Annex B, Table B.13.

3.7 CALCULATION OF STRESS RANGES

3.7.1 Introduction

As in EN 1993-1-9, the authors have separated the calculation of stresses resulting from the action effects (see sub-chapter 3.3 to 3.6) from the calculation of stress ranges. In most situations, the potential fatigue crack will be located in parent material adjacent to some form of stress concentration, e.g. at a weld toe or bolt hole. Provided that the direction of the principal stress does not change significantly in the course of a stress cycle, the relevant cyclic stress for fatigue verification should then be taken as the maximum range through which any principal stress passes in the parent metal adjacent to the potential crack location. All other cases are dealt with in section 3.7.5. Note that in practice the through-thickness component of stress is rarely relevant and can be ignored.

The basis of fatigue and the definition of stress range were given in chapter one, equation (1.2). For simplified fatigue verification with damage correction factors, the fatigue load model ($\gamma_{Ff}Q_k$) is positioned in the two most adverse positions in order to get the maximum stress, σ_{max}, and minimum stress, σ_{min}. The stress range is then computed similarly to expression (1.2) and is given below as expression (3.46).

$$\Delta\sigma_{Ed}\left(\gamma_{Ff}Q_k\right) = \sigma_{Ed,max}\left(\gamma_{Ff}Q_k\right) - \sigma_{Ed,min}\left(\gamma_{Ff}Q_k\right) \qquad (3.46)$$

3.7 CALCULATION OF STRESS RANGES

The same applies to shear stress ranges, expression (3.47):

$$\Delta\tau_{Ed}\left(\gamma_{Ff}Q_k\right) = \tau_{Ed,max}\left(\gamma_{Ff}Q_k\right) - \tau_{Ed,min}\left(\gamma_{Ff}Q_k\right) \tag{3.47}$$

It should be noted that the behaviour under load of some joints, for example misaligned joints or bolted ring connections, may be significantly non-linear, depending on the level of applied stress. Thus, since the principle of superposition does not apply, separate computations may be needed to get the minimum and the maximum stress values.

For welded details, the presence of high tensile residual stresses result in a stress range always corresponding to the algebraic difference between stresses. This is not the case for non-welded details, for which a different relationship is used (see next section). Then, the stress range in bolted joints is dealt with in section 3.7.3. The computation of stress range in welds is given in section 3.7.4. The combination of stress in different directions, or the combination of direct and shear stresses is dealt with in section 3.7.5. Finally, the computation of stress ranges in steel and concrete composite structures is handled in section 3.7.6.

3.7.2 Stress range in non-welded details

For non-welded details, or stress-relieved welded details, with the important exception of bolts (dealt with in section 3.7.3), the assumption can be made that there is no residual or built-in stresses and thus take advantage of the beneficial effects that applied compressive stresses have on fatigue behaviour. Thus, instead of using equation (3.46) for computing the stress range, another relationship is given in EN 1993-1-9. For better clarity, this relationship can be rewritten as follows:

$$\Delta\sigma_{Ed,red} = \sigma_{Ed,max} - \sigma_{Ed,min} \quad \text{when } \sigma_{Ed,min} \geq 0$$

$$\Delta\sigma_{Ed,red} = \sigma_{Ed,max} - 0.6 \cdot \sigma_{Ed,min} \quad \text{when } \sigma_{Ed,min} < 0$$

$$\text{and } \sigma_{Ed,max} \geq 0$$

$$\Delta\sigma_{Ed,red} = 0.6\left(\sigma_{Ed,max} - \sigma_{Ed,min}\right) \quad \text{when } \sigma_{Ed,max} < 0 \tag{3.48}$$

3. DETERMINATION OF STRESSES AND STRESS RANGES

The stress range can thus be reduced by up to 40 % in the case of a detail always in compression. Some authors express it as an "increase" in fatigue strength, which would be in this case increased by a factor 1.67. In the case of shear stress ranges, reduction is not possible, and the applied stress range $\Delta\tau(\gamma_{Ff} Q_k)$ is always computed with equation (3.47).

Figure 3.30 shows the result of using relationships (3.48). As can be seen, it reduces the part of the cycle in compression. In other words, the mean stress influence can be accounted for by using these relationships. Apart from the EN 1993-1-9 mean stress correction factor, other recommendations contain similar rules, namely FKM (2006) and IIW (2009).

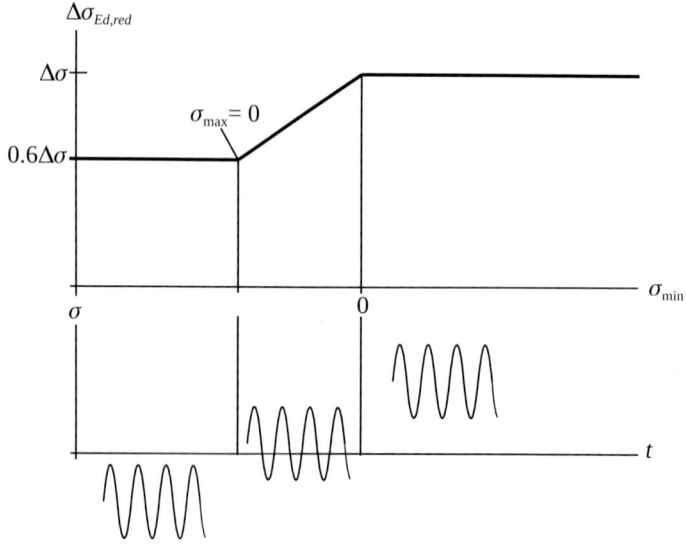

Figure 3.30 – Representation of the reduced stress range for non-welded details

Figure 3.31 shows the mean stress influence expressed in function of the reduced mean stress (that is the value of the mean stress corrected as follows: $\sigma_{mean,red} = \sigma_{Ed,max} - \Delta\sigma_{Ed,red} / 2$, the reduced applied stress range, $\Delta\sigma_{Ed,red}$ and the applied R-ratio. One can see that all recommendations show the same trend, that is the lower the residual stress level, the more the applied stress range is reduced. The reduction is the highest in the case of cycles with part or the full cycles in compression, for negative R-ratios, but in some cases the recommendations also reduce the stress range up to $R = 0.5$.

3.7 CALCULATION OF STRESS RANGES

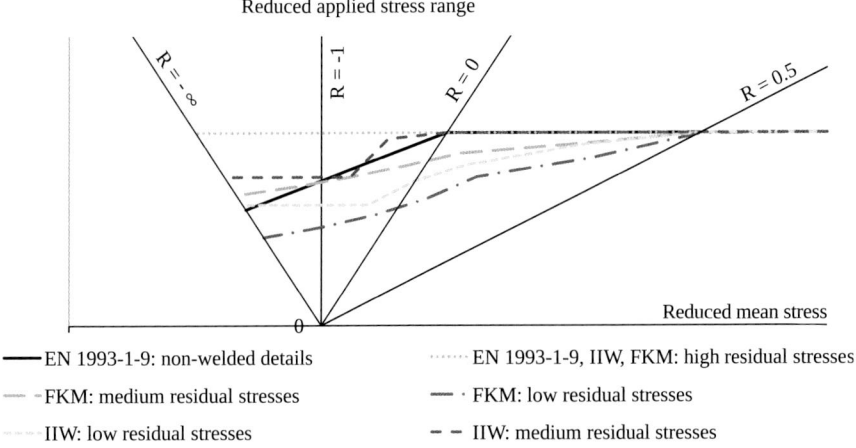

Figure 3.31 – Comparison between different recommendations for computing reduced stress ranges correction factors

3.7.3 Stress ranges in bolted joints

There are three basic load cases bolts can be subjected to:

- bolted connection with preloaded bolts in shear,
- bolted connection with bolts in shear,
- bolted connection with preloaded bolts in tension.

In addition to these basic cases, one can also have bolted connections with preloaded bolts in a combination of shear and tension. One shall emphasize that the fatigue strength of a bolt under tension loading, in opposition with its static strength, is low due to the stress concentrations at the threads. Compared to bolts in shear, bolts in tension represent a much more critical case.

Thus, in the case of bolts under cyclic loading in tension, preloading is a requirement; the use of non-preloaded bolts in tension must absolutely be avoided. This is due to the fact that a non preloaded bolt shank and its threads are subject to stress ranges that are typically an order of magnitude greater than a preloaded one. In a preloaded connection, the bolt preloading acts like a static load, see Figure 3.32. In the connection, there is a state of self-equilibrated, built-in, forces. For the bolt shank, this has the same implications as tensile residual stresses and explains why one cannot benefit from any rule related to mean stress influence to reduce the stress range in the bolt (i.e. the rule for non-welded details given in the previous section).

3. Determination of Stresses and Stress Ranges

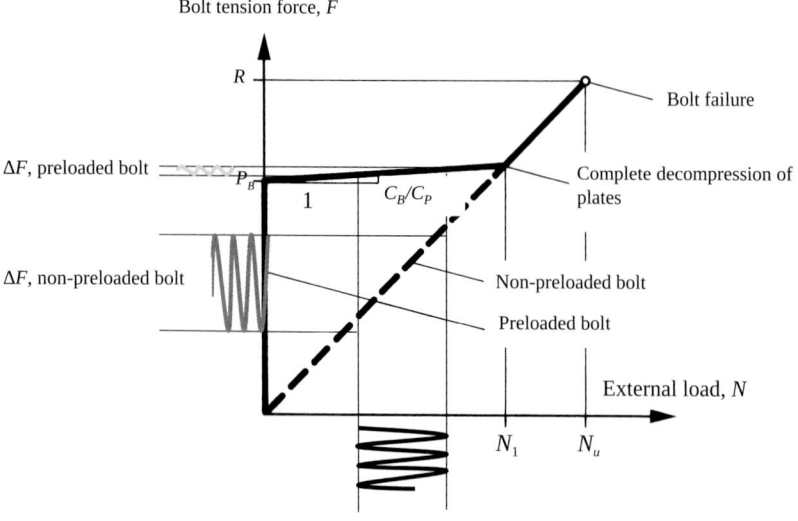

Figure 3.32 – Comparison between the stress ranges in a non-preloaded and a preloaded bolted joints

The initial increased stiffness of a preloaded connection comes from the fact that the connected plate stiffness, C_P, and the bolt axial stiffness, C_B, act together to withstand an external load or load variation ΔN. Indeed, the main part of it goes into decompression of the plates while only a small part goes into the bolts. The load variation in the bolt itself, ΔF, can be globally expressed in function of the stiffness ratio as follows:

$$\Delta F = \Delta N' \cdot \frac{C_B}{C_B + C_P} = p \cdot \Delta N' \qquad (3.49)$$

where
- C_B axial stiffness of the bolt or bolts (based on bolt length, shank area, …),
- C_P axial stiffness of the connection (plate and washers),
- $\Delta N'$ external force range at bolted connection location,
- p distribution factor (to compute the effective force range in the bolt shank).

Since for several cases it is not possible, nor economical, to perform a detailed analysis of the connection behaviour to find the stiffness ratio, one

3.7 CALCULATION OF STRESS RANGES

can use as a first approximation the following value (which is conservative for small prying effects):

$$\Delta F = 0.2 \cdot \Delta N' \qquad (3.50)$$

where $\Delta N'$ is the total external force variation at the bolted connection, including the compression part if any, acting on the connection or the part devoted to one bolt. For bolted connections in tension, a good execution quality is required in order to avoid the effects resulting from imperfections such as eccentricities or lack of good contact between uneven surfaces. These imperfections lead to additional tension as well as bending stresses on the bolts. The method given by Schaumann and Seidel (2001) allows to account for these effects.

In addition, one has to pay a particular attention to the effects of prying forces and to account for any in the stress range computation by using a modified nominal stress range approach. An example with two different connections and the measured resulting prying forces that develop is given in Figure 3.33. To avoid prying forces, two criteria must be considered: good quality of execution and a minimum stiffness of the connected elements.

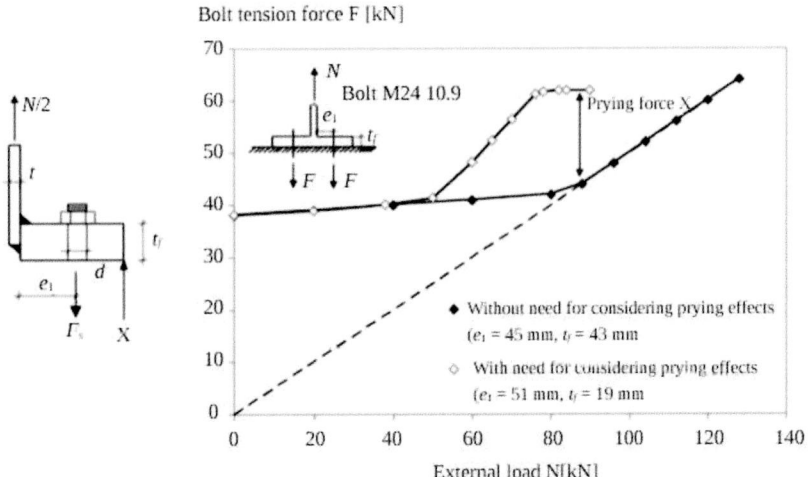

Figure 3.33 – Evolution between external load and bolt tension force for a non-preloaded and a preloaded bolt (NCHRP, 2002)

For the distribution factor, different computations methods are available in the specialized literature, such as Petersen (2000), VDI (2003) or

3. DETERMINATION OF STRESSES AND STRESS RANGES

Schaumann (2001). For a L-joint part of a bolted ring connection in tension, the formulas from Petersen (2000) to compute the various stiffnesses, resulting distribution factor and force range at the bolted connection are given below (see Figure 3.34 for notations) in function of the external force range in the shell ($\Delta N = e \cdot s \cdot \Delta \sigma_{shell}$):

$$\Delta N' = \frac{e \cdot s \cdot \Delta \sigma_{shell} \cdot \left(\frac{b-s}{2}\right) \cdot \left(1 + \frac{a + \frac{s}{2}}{b'}\right)}{\left(\frac{b}{2} + a\right)} \tag{3.51}$$

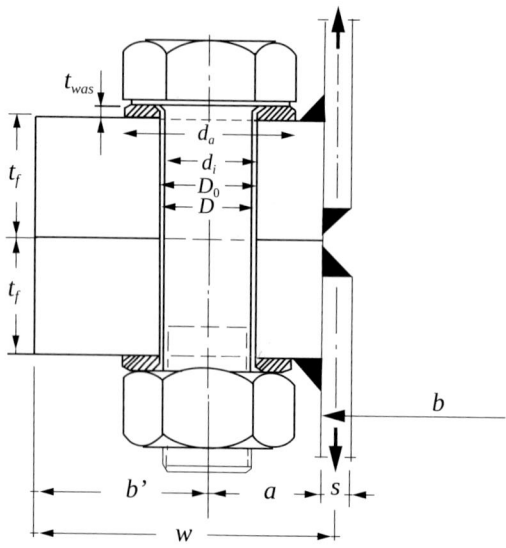

Figure 3.34 – Geometry and notations for L-joint part of a bolted ring connection

Note that the above formula accounts for prying effects. The formulas for bolt axial stiffness, C_B, for flange stiffness, $C_{P,f}$, and for washer stiffness, $C_{P,was}$, are the following (see Figure 3.34 for notations):

$$C_B = \frac{E \cdot \pi \cdot D^2 / 4}{2 \cdot t_f + 2 \cdot t_{was}} \tag{3.52}$$

3.7 CALCULATION OF STRESS RANGES

$$C_{P,f} = \frac{E\,\pi}{2t_f\,4}\left[\left(d_a + \frac{2t_f}{10}\right)^2 - d_i^2\right] \tag{3.53}$$

$$C_{P,was} = \frac{E\,\pi}{t_{was}\,4}\left(d_a^2 - d_i^2\right) \tag{3.54}$$

The resulting connection stiffness, C_P, is then expressed as:

$$C_P = \frac{1}{{}^{1}\!/\!_{C_{P,f}} + {}^{2}\!/\!_{C_{P,was}}} \tag{3.55}$$

Finally the distribution factor p can be computed as follows:

$$p = \frac{C_B}{C_B + C_P} \tag{3.56}$$

Example 3.7: Application to chimney, computation of stress ranges in bolted socket joint (located at + 11490 mm) (worked example 2)

In this part of the example, the fatigue verification of the socket joint is performed. The geometry and dimensions are given in section 1.4.3, and for the socket joint in sub-section 1.4.3.3. The wind loads are computed in section 3.1.5 (Example 3.1).

<u>Loading</u>

Direct stress range in the shell resulting from bending:

$$\Delta\sigma_{shell} = \frac{\Delta M_1}{W_y} = \frac{508.3 \cdot 10^6}{24.493 \cdot 10^6} = 20.8 \text{ N/mm}^2$$

Note: it is interesting to compute the value of the stress due to the dead weight in order to compare it to the fatigue action effects from wind. The axial stress due to dead weight (without partial factors, and considering that the ratio structural to total weight, $W_s/W_t = 0.84$):

$$\sigma_{Gk} = 80000/10^9 \cdot (h - 11500) \cdot 1/0.84 = 4.2 \text{ N/mm}^2$$

3. Determination of Stresses and Stress Ranges

It can be seen that dead weight is only about 21 % of the stress range and thus will not influence fatigue behaviour nor the consequences of failure (the chimney socket joint not being always under compression).

The full bending moment range is to be accounted for since the bolts are pre-tensioned. But the force range acts on the connection, not the bolt alone. In order to know the part of the force range that actually goes into the bolt, an analysis shall be carried out using information from specialized literature, such as Petersen (2000), VDI (2003) or Schaumann and Seidel (2001). The computation below is made using Petersen's method.

Force range in pretensioned bolted connection (for the most loaded bolt)

$$\Delta N' = \frac{e \cdot s \cdot \Delta \sigma_{shell} \cdot \left(\frac{b-s}{2}\right) \cdot \left(1 + \frac{a + \frac{s}{2}}{b'}\right)}{\left(\frac{b}{2} + a\right)} = 56.5 \text{ kN}$$

Bolt stiffness : $C_B = E \cdot A_{bolt} / L_{bolt} = 1350$ kN/mm
Flange stiffness : $C_{P,f} = 5600$ kN/mm
Washer stiffness : $C_{P,was} = 71746$ kN/mm
Flange/washer stiffness : $C_P = 1 / (1/ C_{P,f} + 2/ C_{P,was}) = 4843$ kN/mm
Distribution factor p (force range in one bolt) : $p = C_B/(C_B + C_P) = 0.218$

This means that 78 % of the load range decompresses the connected plates and around 22 % goes into the bolt shank.

Remark: the distribution factor is very close to the first approximation value of 0.2 given in section 3.7.3; prying effects are probably very limited in this case. When using another possible method (VDI, 2003) another value, lower and somewhat unconservative, is found: $p = 0.150$.

Finally, with the value obtained using Petersen (2000), the force range in the most loaded bolt shank is: $\Delta F = \Delta N' \cdot p = 12.3$ kN

Direct stress range in the pre-tensioned bolt:

$$\Delta\sigma_E = \frac{\Delta F}{A_s} = 22.0 \text{ N/mm}^2$$

Number of load cycles (50 years), see section 3.2.6 and Example 3.5.
$N = 8.60 \cdot 10^8$ cycles

Remark: Due to the large number of load cycles ($> 10^8$) the fatigue analysis requires infinite life for all details.

3.7.4 Stress range in welds

This section does not deal with fatigue cracks starting from the weld toe, which is considered outside the weld itself. One differentiates the load carrying welded joints from the other welded joints such as those in welded built-up section or transverse butt welds. The stress ranges in welds are usually computed using nominal stresses (see section 3.3.4).

For transverse butt welds (EN 1993-1-9, Table 8.3), where the fatigue crack might start either in the weld or from the weld toe, the stress range is computed in the adjacent base metal as the area of the weld shall be equal to or larger than the area of the attached elements.

For the welds such as those in welded built-up section, the relevant nominal stress range in the parent material or in the section at the position of the weld, e.g. for longitudinal welds, shall be calculated.

In load-carrying partial penetration or fillet-welded joints, where cracking could occur in the weld throat, one must compute separately the following stress ranges corresponding to the two cases shown in Figure 3.35. In this case, a total of three separate fatigue checks must be carried out (i.e. with three different detail categories). The three cases are listed below:

- Nominal normal stress range in the weld (linked with case A), $\Delta\sigma_w$, to check against fatigue category 36*
- Nominal shear stress range in the weld (linked with case A), $\Delta\tau_w$, to check against fatigue category 80 (m =5)
- Nominal direct stress range at weld toe calculated in the plate (linked with case B), $\Delta\sigma$, to check against fatigue categories for detail 1, Table 8.5 (which is a function of attachment thickness, t, and total attachment length, L)

3. Determination of Stresses and Stress Ranges

One shall repeat here that since tensile residual stresses are assumed to exist in all welded joints, none of the load is carried in bearing between parent materials (in the gap), even if the joint is under compressive loads.

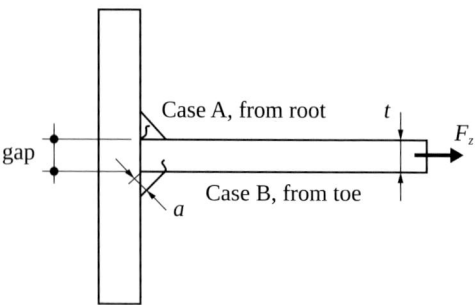

Figure 3.35 – Possible fatigue crack locations in partial penetration a T-joints, butt or fillet welded (double-sided)

Example 3.8: Application to chimney, stress range in the welded stiffener on the bottom ground plate (located at +350 mm) (worked example 2)

The geometry and dimensions are given in section 1.4.3, and for the assembly at the bottom in section 1.4.3.4. The wind loads are computed in section 3.1.5 (Example 3.1).

Loading

Force range in the longitudinal stiffener (computed according to the simplified method from Petersen):

$$\Delta F = \frac{2}{n} \cdot \frac{\Delta M_4}{r_s} = 49.4 \text{ kN}$$

Stress range in throat of weld between stiffener and ground plate (with $\tau_\perp = 0$), from expressions (3.26) and (3.27):

Case A:

$$\sigma_{E,w} = \sqrt{\left(\frac{F_z}{2a_{eff} \cdot \ell}\right)^2 + \tau_\parallel^2}$$

$$\Rightarrow \Delta\sigma_{E,w} = \frac{\Delta F}{2(a_w+1) \cdot L_w} = \frac{49400}{2 \cdot 7 \cdot 220} = 16.0 \text{ N/mm}^2$$

$$\tau_w = \tau_{//} = \frac{F_y}{2a_w \cdot L_w} = 0$$

Case B (expression (3.28)):

$$\sigma = \frac{F_z}{t \cdot L_w} \Rightarrow \Delta\sigma = \frac{\Delta F_z}{t \cdot L_w} = \frac{49400}{12 \cdot 220} = 18.7 \ \text{N}/\text{mm}^2$$

3.7.5 Multiaxial stress range cases

3.7.5.1 Introduction

The particular case of stress ranges in a weld, where the normal stress range (based upon the direct stress ranges acting on the detail) and shear stress ranges are computed separately and not combined has been seen in the previous section. More generally, when normal and shear stresses are likely to cause the formation of fatigue cracks in the same detail but at distinct locations, as for example in lap joints, see Table 8.5, details 5 and 9, a separate verification for both locations should be performed. In these cases, no combination is needed as it is specified for example in BS7608 (1993). Note that in the sections related to multiaxial stresses, the generic term normal stress will used indifferently for both the stresses acting on a detail or in a weld.

All other cases, when combination is needed, configure the general case of details under multiaxial stress ranges which is now dealt with. A single loading or multiaxial cyclic loading results in different stress components which can be any combination of either normal and/or shear stresses. The different common cases are explained below. The question of the fatigue verification under multiaxial stress ranges is not treated here, but later in section 5.4.7. Multiaxial stress problems can be dealt with both using a nominal stress or geometric stress approach but, in this manual, the emphasis will be made on nominal stress, even if some of the concepts and formulas are also applicable to a geometric stress approach. Furthermore, note that alternative methods such as Dong's mesh-insensitive, structural stress method (Dong *et al*, 2006) are also able to deal with fatigue design under multiaxial stresses.

In the subsequent explanations, reference is made to the principal stresses and directions of the stress tensor, not to the mean hydrostatic pressure and the stress deviator tensor which is also used in damage mechanics.

3.7.5.2 Possible stress range cases

The different possible cases, from the simplest to the more complex, are described below, adapted from FKM (2006), and also shown in Figure 3.36, adapted from Radaj (2003):

- **Proportional stresses:** they usually result from a single loading, varying with time, acting on the structural member. All multiaxial stresses are varying proportionally to that loading and proportionally to each other, which is also true with regard to their ranges and their mean values. Further, as a consequence, the principal stress directions stay constant; all the same for the principal directions of the stress ranges. Examples of proportional stresses are the circumferential and the longitudinal stresses of a cylindrical vessel loaded by internal pressure or the bending and torsion moment stresses of a cantilevered mast with an asymmetric arm and, as a consequence, loaded eccentrically by wind loads.
- **Non-proportional, synchronous stresses (or in-phase stresses):** this is the simplest case of non-proportional stresses since the loadings are in-phase (however can be in opposition, 180° shift) and only non-proportional with regard to their mean values. Synchronous stresses usually result from the combined action of a constant load with a second, different kind of loading, varying with time. Thus, the resulting stress ranges are proportional, i.e. if the varying loading doubles then each of the multiaxial stress range components double, but not their mean values. As a consequence, the principal stress directions change (however only if one forces the first principal stress to be the maximum) but not the principal directions of the stress ranges. Note that the stress component waveforms resulting from the loading may also be different. Examples are a shaft with a non-changing torsion moment together with a rotating bending loading; or a long, lying cylindrical vessel under pulsating internal pressure, where the longitudinal stress is non-proportional to the circumferential stress (because the bending stress from the dead load is additively overlaid to the pressure stresses).

3.7 CALCULATION OF STRESS RANGES

- **Non-proportional, asynchronous stresses:** all multiaxial stresses that are not synchronous are called asynchronous stresses; they can result from out-of-phase loadings or from loadings with different frequencies.
- **Non-proportional stresses:** they usually result from the action of at least two loadings that vary non-proportionally with time in a different manner. They can however also result from one constant combined with one moving load. Thus they may result in synchronous or asynchronous stresses. In the most general case of non-proportional loading, i.e. variable amplitude stress components histories with different frequencies, different spectra apply to the individual stress components that result from the combined loadings. As a consequence, both the principal stress directions and the principal directions of the stress ranges change with time. In addition, the time reference points that correspond to the minimum and maximum of a stress component (and thus to the maximum stress range for this component) may be different.

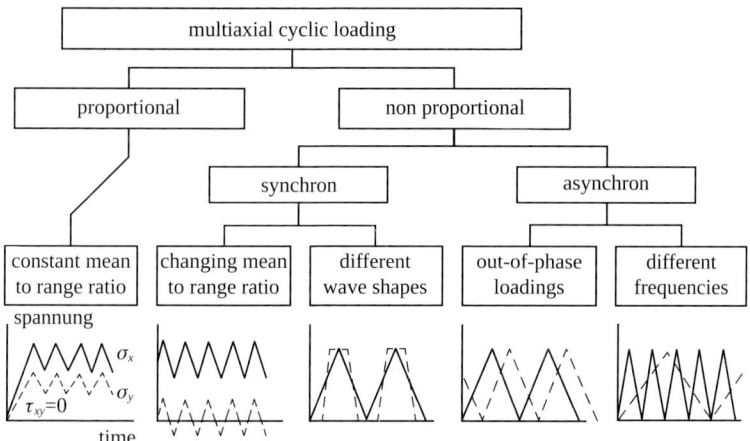

Figure 3.36 – Differentiation between proportional and non-proportional cyclic loadings and further separation of the different cases (Radaj, 2003)

3.7.5.3 Proportional and non-proportional normal stress ranges

In this particular case, by definition, the stress components directions correspond to the principal directions. As a consequence, for both proportional and non-proportional loadings, the principal directions of the stress ranges stay the same. Then, both the principal normal stress (BS7608, 1993) or the von Mises stress (EN1993-1-9, 2005) can be

analysis of the composite steel and concrete member. Several effects cannot be neglected in the analysis for fatigue limit state verifications, namely:

- primary and secondary effects caused by shrinkage and creep of the concrete flange
- the effects of sequence of construction
- temperature effects.

Furthermore, for fatigue verifications, one cannot use rigid plastic global analysis, nor elastic bending moments redistribution in continuous beams.

In buildings, the generic relevant Eurocode parts do not require fatigue verifications except in very specific cases. The relevant clauses are the following:

- For concrete, EN 1992-1-1, clauses 6.8.1(1) and (2). Fatigue verifications should be carried out only for structures and structural components which are subjected to regular load cycles (e.g. crane-rails, bridges exposed to high traffic loads). In these cases, the verification shall be performed separately for concrete and reinforcing steel.
- For structural steel, EN 1993-1-1, clause 4(4). Exceptions are members supporting lifting appliances or rolling loads, subjected to repeated stress cycles from vibrating machinery, subjected to wind-induced vibrations or to crowd-induced oscillations.
- For steel-concrete composite structures, EN 1994-1-1, clause 6.8.1(4). No fatigue assessment for structural steel, reinforcement, concrete and shear connection is required where, for structural steel, EN 1993-1-1, clause 4(4) applies and, for concrete EN 1992-1-1, clause 6.8.1, does not apply.

Since bridges is the main domain for which fatigue verifications in steel and concrete structures are applied, this section presents in a detailed manner the method for bridges (the method shown also being applicable to other types of steel and concrete composite structures).

Consider a steel-concrete composite bridge. According to EN 1994-2, clause 6.8, $M_{Ed,max,f}$, respectively $M_{Ed,min,f}$, is the maximum, respectively the minimum, value of the bending moment coming from one of the combination of actions defined in expressions (3.31) and (3.32) of this Manual. With consideration for the different construction phases, it can be expressed for example for the maximum value as:

3.7 CALCULATION OF STRESS RANGES

$$M_{Ed,\max,f} = M_{a,Ed} + M_{c,Ed} + M_{FLM3,\max} \qquad (3.58)$$

where

$M_{a,Ed}$ design value of the bending moment from basic SLS combination of non-cyclic loads (see EN 1992-1-1, clause 6.8) applied to the structural steel section before connexion (composite behaviour),

$M_{c,Ed}$ design value of the bending moment from basic SLS combination of non-cyclic loads (see EN 1992-1-1, clause 6.8) applied to the steel-concrete section after connexion (composite behaviour),

$M_{FLM3,\max}$ maximum bending moment due to fatigue load model FLM3, see sub-section 3.1.2.3.

Three different situations are then considered for the stress range calculation as follows:

1) $M_{Ed,\max,f}$ and $M_{Ed,\min,f}$ cause tensile stresses in the concrete slab.

The effect of the basic SLS combination for non-cyclic loads disappears from the stress range, which should be calculated using the mechanical properties of the composite cross section with cracked concrete (structural steel + reinforcement) :

$$\sigma_{\max,f} - \sigma_{\min,f} = \left(M_{FLM3,\max} - M_{FLM3,\min}\right)\frac{v_2}{I_2} \qquad (3.59)$$

where

v_2 distance from the neutral axis to the relevant fibre,
I_2 inertia of the cracked concrete composite cross section,
M_{FLM3} bending moment (minimum or maximum) due to fatigue load model FLM3, see sub-section 3.1.2.3.

2) $M_{Ed,\max,f}$ and $M_{Ed,\min,f}$ cause compression stresses in the concrete slab.

The effect of the basic SLS combination for non-cyclic loads also disappears from the stress range, which should be calculated using the mechanical properties of the composite cross section with uncracked concrete (structural steel + concrete):

3. DETERMINATION OF STRESSES AND STRESS RANGES

$$\sigma_{max,f} - \sigma_{min,f} = \left(M_{FLM3,max} - M_{FLM3,min}\right)\frac{v_1}{I_1} \quad (3.60)$$

where v_1 is the distance from the neutral axis to the relevant fibre and I_1 is the inertia of the uncracked concrete composite cross section, calculated with the short term modular ratio $n_0 = E_a / E_{cm}$.

3) $M_{Ed,max,f}$ causes tensile stresses and $M_{Ed,min,f}$ causes compression stresses in the concrete slab.

In this situation, the composite part of the bending moment from the basic SLS combination for non-cyclic loads, $M_{c,Ed}$, influences the stress range according to the following equation:

$$\sigma_{max,f} - \sigma_{min,f} = M_{c,Ed}\left[\frac{v_2}{I_2} - \frac{v_1}{I_1}\right] + M_{FLM3,max}\frac{v_2}{I_2} - M_{FLM3,min}\frac{v_1}{I_1} \quad (3.61)$$

$M_{c,Ed}$ is normally split up into several action effect cases for which the corresponding stresses should be evaluated with the proper elastic modular ratio n_L. In order to simplify the calculations, the short-term elastic modulus ratio n_0 may also be used for all the action effects.

Example 3.9: Application to steel and concrete composite road bridge, computation of stress ranges (worked example 1)

The elastic global cracked analysis of a composite bridge is dealt with in EN 1994-2, clause 6.8. The bending moment for the basic SLS combination of non cyclic loads is given below with minimum explanations; more details can be found in SETRA (2007).The box-girder bridge has two slow lanes located on the right-hand side of each direction, respecting the painted marks (see Figure 1.13). For bending moment computations, the construction phasing of the deck (concreting of the slab in segments) has been taken into account. The results of the computations are given in Figure 3.38.

3.7 CALCULATION OF STRESS RANGES

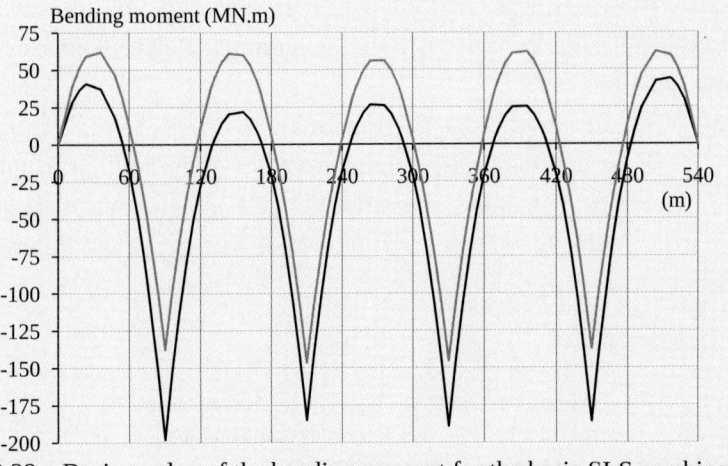

Figure 3.38 – Design value of the bending moment for the basic SLS combination of non-cyclic loads

In each cross section, two extreme values are obtained for the basic SLS combination $M_{a,Ed} + M_{c,Ed}$. Then, the FLM3 passage is added to each of these extreme values as follows:

$$\begin{cases} M_{Ed,max,f} = \min(M_{a,Ed} + M_{c,Ed}) + M_{FLM3,max} \\ M_{Ed,min,f} = \min(M_{a,Ed} + M_{c,Ed}) + M_{FLM3,min} \end{cases}$$

$$\begin{cases} M_{Ed,max,f} = \max(M_{a,Ed} + M_{c,Ed}) + M_{FLM3,max} \\ M_{Ed,min,f} = \max(M_{a,Ed} + M_{c,Ed}) + M_{FLM3,min} \end{cases}$$

In each cross section and for each fibre, a corresponding stress range $\Delta\sigma = |\sigma_{max,f} - \sigma_{min,f}|$ can then be calculated for both extreme values. As an example Figure 3.39 illustrates the obtained stress ranges on the lower face of the upper flange of the box girder. Each peak value above the envelope of the stress range resulting from FLM3 crossing illustrates the influence of the basic SLS combination of non-cyclic loads, i.e. meaning that $M_{Ed,max,f}$ causes tensile stresses and $M_{Ed,min,f}$ causes compression stresses in the concrete slab. It results in high stress range values at quarter span cross sections. According to the method from EN 1994-2, quarter span regions become often critical from the point of view of fatigue and should be carefully checked. These peaks may be avoided by optimisation of the construction phasing of the deck and by careful choice of fatigue details and their location.

In Figure 3.39, the stress range corresponding to expression (3.31), case 1 in section 3.3.5, is mentioned as coming from the minimum SLS non-cyclic load combination of actions. The stress range corresponding to expression (3.31), case 2, comes from the maximum SLS non-cyclic load combination of actions.

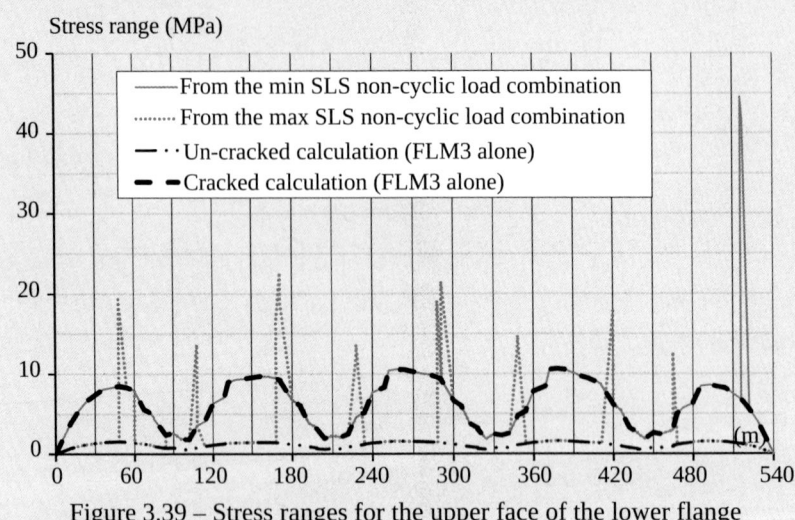

Figure 3.39 – Stress ranges for the upper face of the lower flange

3.7.7 Stress ranges in connection devices from steel and concrete composite structures

In a steel and concrete composite structure subject to fatigue loadings, one important issue is the fatigue verification of the connection. In this specific case, the stresses acting in the detail are:

– a direct stress range in the steel beam flange, to which the stud connectors are welded,
– a shear stress range in the weld of each of the stud connectors due to the composite action effect between the concrete slab and the steel beam.

It should be first noticed that EN 1994-2 only deals with welded studs.

3.7 CALCULATION OF STRESS RANGES

Other types of shear connector (angles, ...) have to be introduced in the National Annexes if necessary, see EN 1994-2, clause 1.1.3(3).

The method for determining the direct stress range in the flange has been explained in the previous section. The method for determining the shear stress and shear stress ranges is now presented; it is made according to EN 1994-2, clauses 6.8.5.5 and 6.8.6.2. The shear stresses at the steel-concrete interface are calculated using the properties of the cross section with uncracked concrete (in opposition to the direct stress calculations). As a consequence, the basis SLS combination of non-cyclic loads has no influence on the shear stress range, which is only induced by the FLM3 crossing and computed, as usual, as the difference between the two extreme values.

The longitudinal shear force per unit length is computed as follows:

$$v_L = \frac{S_{V1} \cdot V_{Ed}}{I_1} \qquad (3.62)$$

where
- V_{Ed} design value of the longitudinal shear force computed from a global cracked concrete analysis
- S_{V1} first moment of area of the concrete slab (taking the shear lag effect into account by means of an effective width) with respect to the centroid of the uncracked composite cross section
- I_1 second moment of area of the uncracked concrete composite cross section.

The expression for the shear stress range is:

$$\Delta\tau = \frac{\Delta v_{L,FLM3}}{A_{stud} \cdot n_{stud}} \qquad (3.63)$$

where
- $\Delta v_{L,FLM3}$ longitudinal shear force per unit length at the steel-concrete interface due to FLM3 crossing
- A_{stud} shear area of a connector
- n_{stud} number of shear studs per unit length.

3. DETERMINATION OF STRESSES AND STRESS RANGES

Example 3.10: Application to steel and concrete composite road bridge, computation of shear stress ranges (worked example 1).

In order to focus on the fatigue verifications in shear connection, the shear connection density (number of connectors per unit length) is given here as a starting hypothesis (no detailed calculations are provided since it is out of the scope of this Design Manual). It has been determined according to the method given in EN 1994-2, clause 6.6. As a result, the following choice for the connection has been made:

- stud diameter: $d = 22$ mm
- stud height: $h = 200$ mm (to be sufficiently anchored in the reinforced concrete slab)
- 4 studs in a transversal row for each steel upper flange of the box-girder ultimate limit strength for the steel of the studs : $f_u = 450$ MPa
- the connection density is constant over a given section of the bridge length, the section cutting depending on the SLS and ULS shear flow distributions, see Figure 3.40
- the following recommended values have been adopted for the design resistance of a single stud: $\gamma_V = 1.25$ (see EN 1994-2, clause 6.6.3.1) and $k_s = 0.75$ (see EN 1994-2, clause 6.8.1).

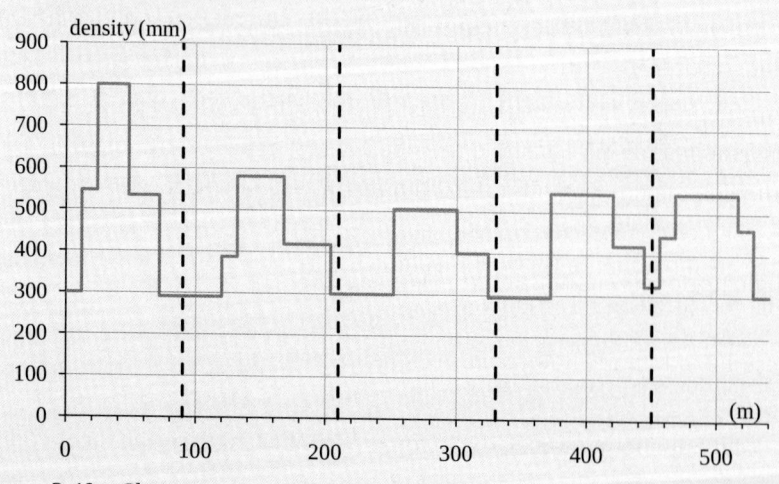

Figure 3.40 – Shear connectors row spacing (density) for one of the steel flanges

3.7 CALCULATION OF STRESS RANGES

The shear area of a connector is equal to $A_{stud} = \dfrac{\pi d^2}{4} = 380.1 \text{ mm}^2$.

For a given cross section, for example one located near the mid-span P1-P2 at $x = 156$ m, the detailed calculations give $I_1 = 2.2 \text{ m}^4$ for half of the bridge cross section, with:

$$n_0 = \frac{E_a}{E_{cm}} = \frac{E_a}{22000\left(\dfrac{f_{ck}+8}{10}\right)^{0.3}} = \frac{210000}{22000 \cdot \left(\dfrac{35+8}{10}\right)^{0.3}} = 6.1625$$

for the short-term modular ratio. The centroid of the un-cracked cross section is located in the main steel web, 675 mm below the steel concrete interface, and the width of the concrete slab is equal to $21.5/2 = 10.75$ m (no shear lag effect at this location) for a slab thickness of 325 mm.

The first moment of area is given by:

$$S_{V1} = 10.75 \cdot (0.325/6.1625) \cdot (0.675 + 0.325/2) = 0.475 \text{ m}^3$$

In this cross section, when FLM3 crosses the bridge, a shear force range is created, given by:

$$\Delta V_{FLM3} = V_{max,FLM3} - V_{min,FLM3} = 97.2 - (-120.8) = 218 \text{ kN}.$$

Then, the shear flow per unit length is equal to (expression (3.62)):

$$\Delta v_{L,FLM3} = \frac{\Delta V_{FLM3} S_{V1}}{I_1} = \frac{218 \cdot 0.475}{2.2} = 47.1 \text{ kN/m}.$$

Finally, with a 4-studs row spacing of 580 mm between $x = 132$ m and $x = 168$ m, the shear stress range is computed using expression (3.63) as:

$$\Delta \tau = \frac{\Delta v_{L,FLM3}}{A_{stud} n_{stud}} = (47.1 \cdot 10^{-3})/(380.1 \cdot 10^{-6})/(4/0.580) = 17.9 \text{ MPa}$$

Figure 3.41 illustrates the variation of this shear stress range along the entire bridge, with an indication at the cross section location $x = 156$ m corresponding to the computation detailed in this example.

3. DETERMINATION OF STRESSES AND STRESS RANGES

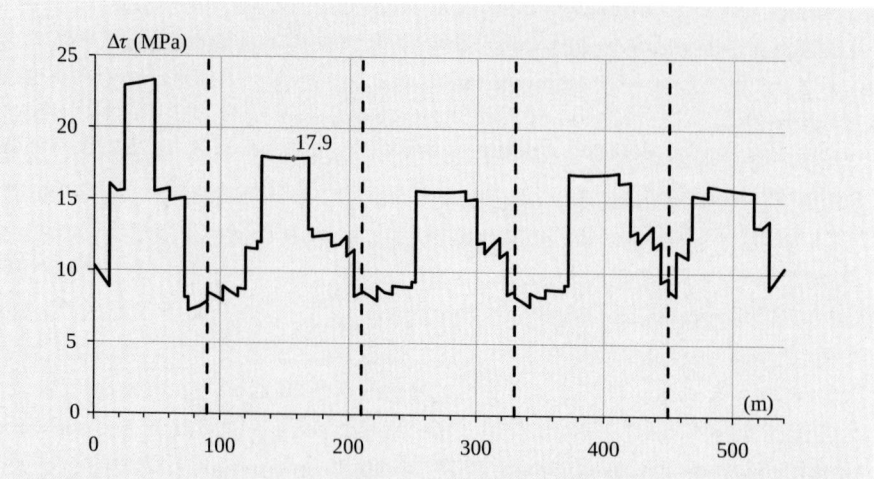

Figure 3.41 – Shear stress ranges at the steel-concrete interface along one of the steel upper flange

3.8 MODIFIED NOMINAL STRESS RANGES

In analogy to sub-chapter 3.4, formula (3.33), the expression for the modified nominal stress range is in general:

$$\Delta\sigma_{mod} = k_f \cdot \Delta\sigma_{nom} \quad (3.64)$$

where k_f is the geometric stress concentration factor to account for the local stress magnification in relation to detail geometry not included in the reference S-N curve. The values for k_f are to be taken from handbooks or from appropriate finite elements calculations as explained in sub-chapter 3.4. Expression (3.64) may not be appropriate for cases where the behaviour under load of the detail is highly non-linear.

In EN 1993-1-9, the relationship given for computing the modified stress range directly includes the damage equivalent factor. Explicitly putting it into the expression leads to the following relationship

$$\gamma_{Ff} \cdot \Delta\sigma_{mod,E,2} = \lambda \cdot k_f \cdot \Delta\sigma_{nom}\left(\gamma_{Ff} \cdot Q_k\right) \quad (3.65)$$

3.8 MODIFIED NOMINAL STRESS RANGES

Additionally to the geometric stress concentration in a classified structural detail configuration and the geometric stress concentration accounted for using k_f, other cases such as misalignment can occur. The resulting additional stresses reduce the fatigue strength and have to be accounted for. In Eurocode 3, they are included through the multiplication by a factor k_s that reduces the fatigue strength (expression 3.66):

$$\Delta\sigma_{C,red} = k_s \cdot \Delta\sigma_C \qquad (3.66)$$

k_s reduction factor for fatigue strength to account for size effects and/or eccentricity. Note that $1/k_f = k_s$

Such a case is found in the classification tables for detail 17 of Table 8.3 (see Figure 3.42).

$\Delta\sigma_c$	Constructional detail		Description
71	Size effect for $t > 25$ mm and/or generalisation for eccentricity: $k_s = \left(\dfrac{25}{t_1}\right)^{0.2} \Big/ \left(1 + \dfrac{6e}{t_1} \dfrac{t_1^{1.5}}{t_1^{1.5} + t_2^{1.5}}\right)$	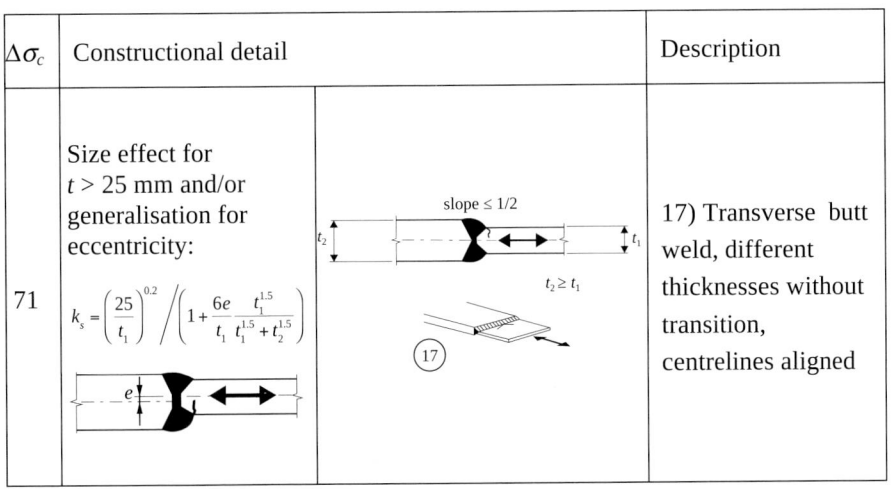	17) Transverse butt weld, different thicknesses without transition, centrelines aligned

Figure 3.42 – Example of detail with additional geometric stress concentration due to misalignment, or say eccentricity

This case is a good example because it is particular as two different influences are combined in a single factor. These different influences can be separated as follows:

1. Geometric concentration factor (multiplying action effects, $\Delta\sigma(\gamma_{Ff}Q_k)$, or as here reducing strength) for misalignment (see also sub-chapter 3.4, expression (3.36))

3. DETERMINATION OF STRESSES AND STRESS RANGES

$$k_{s,1} = \left(1 + \frac{6e}{t_1} \frac{t_1^{1.5}}{t_1^{1.5} + t_2^{1.5}}\right)^{-1} \quad (3.67)$$

2. Size factor (reducing strength, $\Delta\sigma_C$), see section 4.1.3:

$$k_{s,2} = \left(\frac{25}{t_1}\right)^{0.2} \quad (3.68)$$

The way it is done in the Eurocode is not limited to the above case, which can be confusing since some geometric stress concentration, or stress increasing effects, are taken on the action side as multipliers and some others on the resistance side as dividers.

In the case of welded joints of hollow sections, expression (3.65) is rewritten in the Eurocode as follows:

$$\gamma_{Ff} \cdot \Delta\sigma_{hs,E,2} = \lambda \cdot k_1 \cdot \Delta\sigma^*_{E,2}\left(\gamma_{Ff} \cdot Q_k\right) \quad (3.69)$$

where

$\gamma_{Ff} \cdot \Delta\sigma_{hs,E,2}$ design value of the modified nominal stress range

k_1 stress magnification factor to account for secondary bending, see section 3.3.6, method 1 (referring to Tables 4.1 and 4.2 of EN 1993-1-9) to account for secondary bending, see section 3.3.6

$\Delta\sigma^*_{E,2}\left(\gamma_{Ff} \cdot Q_k\right)$ design value of the nominal stress range calculated with the method of analysis 1 or 2a simplified truss model with pinned joints as explained in section 3.3.6.

3.9 GEOMETRIC STRESS RANGES

The determination of the structural stress at the hot spot range follows directly from computing the difference between the maximum and minimum geometrical stress values. The methods to compute geometric stresses were presented in sub-chapter 3.5. In EN 1993-1-9, the relationship given for computing the geometric stress range directly includes the damage

3.9 GEOMETRIC STRESS RANGES

equivalent factor. Explicitly putting it into the expression leads to the following relationship:

$$\gamma_{Ff} \cdot \Delta\sigma_{hs,E,2} = \lambda \cdot k_f \cdot \Delta\sigma_{nom}\left(\gamma_{Ff} \cdot Q_k\right) \tag{3.70}$$

where

$\gamma_{Ff} \cdot \Delta\sigma_{hs,E,2}$ design value of the structural stress at the hot spot range

k_f geometric stress concentration factor

$\Delta\sigma_{nom}\left(\gamma_{Ff} \cdot Q_k\right)$ design value of the nominal stress range calculated with a simplified model

Geometric stress concentration factor values can be found in the literature for a large number of different details. These factors are obtained analytically or from parametric studies as presented in more detail in sub-chapter 3.5. Formula (3.70) may not be appropriate for cases where the behaviour under load of the detail is highly non-linear.

Example 3.11: Application to a welded tubular truss in an industrial building

This example is not part of the worked examples introduced in chapter one. It deals with a case of geometric stress approach and verification.

An intermediate floor for machinery, with a large span of 36 m, is supported by regulary spaced uniplanar welded tubular trusses, as shown in Figure 3.43, reference CIDECT (2001) and Stahlbaukalender (2006). The truss is made out of circular hollow sections (CHS). The fatigue design values of the applied loading can be modelled as a constant amplitude loading acting at the nodes of the top chord, with values ranging from zero to the loads shown in Figure 3.43.

The member sizes are:

Top chord: CHS 219.1 · 7.1, $A_0 = 4728$ mm², $W_0 = 0.243 \cdot 10^6$ mm³
Braces: CHS 88.9 · 4.0, $A_{1,2} = 1070$ mm², $W_{1,2} = 0.0217 \cdot 10^6$ mm³
Bottom chord CHS 177.8 · 7.1, $A_0 = 3807$ mm², $W_0 = 0.156 \cdot 10^6$ mm³

3. DETERMINATION OF STRESSES AND STRESS RANGES

The joint eccentricities, e, are equal to zero and the static strength has been found satisfactory.

The purpose of this example is to determine the fatigue life of joint 6, according to both the **modified nominal stress** and the **geometric stress method**.

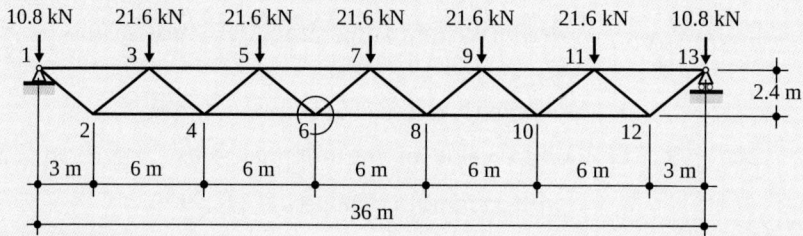

Figure 3.43 – Uniplanar girder and design load situation (constant amplitude load ranges)

Method A: Design according to modified nominal stress method.
The verification is separated in different steps for clarity.

Step 1: Structural analysis

A structural analysis is carried out assuming a continuous chord and pin-ended braces. The axial forces and bending moments found in joint 6 are given in Figure 3.44. They can be treated as a combination of two load conditions shown in Figure 3.45, i.e.:

Load condition 1: basic balanced axial loading
Load condition 2: chord loading (axial and bending)

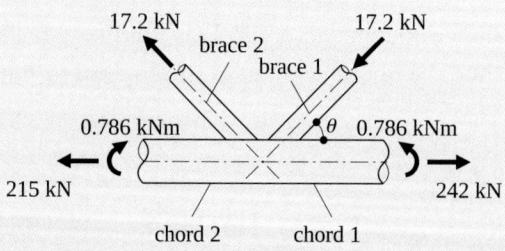

Figure 3.44 – Internal load condition of joint 6 (axial forces and bending moments ranges)

3.9 GEOMETRIC STRESS RANGES

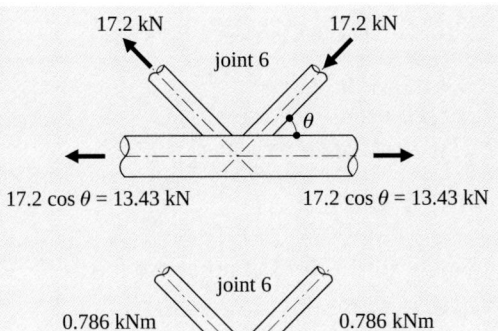

Figure 3.45 – Two basic fatigue load-cases of joint 6

Step 2: Nominal stress ranges in critical members

It can be seen from Figure 3.44 that the critical chord loading is in chord 1 due to a larger tension force. Only brace 2 with a tensile force range will be checked. Note: In general, the braces which have some parts of their load range in tension are usually the ones responsible for fatigue failure (even if braces in tension and in compression will initiate fatigue cracks). The indices are: *br* for brace, *ch* for chord, and then *ax* for axial loading and *ch* for chord loading.

For load condition 1 (basic balanced axial loading):

$$\sigma_{br,ax} = 17.2 \cdot 10^3 / 1070 = 16 \text{ MPa}$$

For load condition 2 (chord loading):

$$\sigma_{ch,ch} = 228.5 \cdot 10^3 / 3807 - 0.786 \cdot 10^6 / (0.156 \cdot 10^6) = 60 - 5 = 55 \text{ MPa}$$

(Note that the chord bending moment relieves the tensile stress on the connecting face of the chord.).

Step 3: Secondary bending moments and stress range values for design

Since the method of analysis of the structure was method 1 or 2 (equivalent in this case since there are no excentricities at the joints), see section 3.3.6, the nominal stress ranges found in step 3 have to be multiplied by magnification factors k_1, to take into account the effects of the secondary bending moments. These factors are given in Table 4.1 of EN 1993-1-9. In this example, the magnification factor values $k_1 = 1.3$ for the braces and $k_1 = 1.5$ for the chord shall be used.

3. DETERMINATION OF STRESSES AND STRESS RANGES

For load condition 1 (basic balanced axial loading):

$$\sigma_{br,ax} = 1.3 \cdot 16 = 21 \text{ MPa}$$

For load condition 2: (chord loading):

$$\sigma_{ch,ch} = 1.5 \cdot 55 = 83 \text{ MPa}$$

Finally, the design values of the stress ranges are thus:

$$\Delta\sigma_{br,ax} = 21 \text{ MPa}$$

$$\Delta\sigma_{ch,ch} = 83 \text{ MPa}$$

Step 4: Determination of Detail category

The detail category can be determined from Table 8.7 of EN 1993-1-9 (table for welded tubular joints, see Annex B.7). The detail category depends upon the loading and geometry of the joint, and the joint has to be within the validity range of the table and of the corresponding detail category.

The maximum thickness of the tubes composing the joint is t_0 = 7.1 mm (chord); this is less than 12.5 mm and thus Table 8.7 can be used. Computation of the geometric parameters of the joint:

$$\beta = d_1 / d_0 = 88.9/177.8 = 0.5$$

$$2\gamma = d_0 / t_0 = 177.8/7.1 = 25$$

$$\gamma = 12.5$$

$$\tau = t_1 / t_0 = 4/7.1 = 0.563$$

$$\theta = \arctan(2.4/3.0) = 38.7°$$

$$e = 0$$

All the requirements listed in Table 8.7 have to be checked and in this case they all are fulfilled. With a thickness ratio of t_0 / t_1 = 1.775, it results in using a detail category 45 (m = 5). A conservative assumption, linear interpolation between 45 and 90 could be used.

Step 5: Determination of partial resistance factor

A partial factor on stress ranges is required for design. For this example the joint is assumed to be damage tolerant (more than one truss supporting the floor, semi-rigid truss joints, crack sites well defined and regular inspection

3.9 Geometric Stress Ranges

possible) and with high consequences for a failure (possibilities of partial collapse on people and high financial losses due to shut down of unit). From Table 3.1 of EN 1993-1-9, the partial factor $\gamma_{Mf} = 1.15$ has been taken.

Step 6: Estimation of the fatigue life for joint 6

Using the higher value of step 3 $\Delta\sigma_{ch,ch} = 83$ N/mm², the fatigue life can be determined either from Figure 4.2 of EN 1993-1-9, or more accurately, using the detail category. Conservatively, the detail category ca be taken as 45, but one is also allowed to extrapolate between categories 45 and 90 since we have a t_0/t_1 inferior to 2 but close, namely 1.775.

$$N_R = 2 \cdot 10^6 \cdot \frac{\left(\Delta\sigma_C / \gamma_{Mf}\right)^m}{\Delta\sigma_{ch,ch}^m} = 2 \cdot 10^6 \cdot \frac{(45/1.15)^5}{83^5} = 46600 \text{ cycles}$$

The fatigue life expectancy of joint 6 is about 46 000 cycles, with fatigue cracking and failure in the chord.

Design according to the geometric stress method (with the help of the CIDECT design guide 8 (CIDECT, 2001))

For tubular structures, the use of the fatigue verification according to the classification method is often not possible because of the limited validity range of Table 8.7, EN 1993-1-9. This is often the case for large tubular structures, large spans and bridges. Furthermore, the detail categories in nominal stress range are generally conservative since they cannot account for the differences in local geometry within the joints. For these reasons and for comparison, the verification of the same joint is now made according to section 6.5 of EN 1993-1-9 and the CIDECT design guide 8.

To simplify, to account for secondary bending in the joints, the values found previously for the modified nominal stress range are used herein (to avoid this simplification and get more precise values, the use of a model as explained in section 3.3.6, method of analysis 3, is also possible). Thus, in this example, *step 1 to step 4 remain unchanged. Recall:*

- The geometric parameters of the joint:

$\beta = d_1 / d_0 = 88.9/177.8 = 0.5$

$2\gamma = d_0 / t_0 = 177.8/7.1 = 25$

3. Determination of Stresses and Stress Ranges

$\gamma = 12.5$

$\tau = t_1 / t_0 = 4/7.1 = 0.563$

$\theta = \arctan(2.4/3.0) = 38.7°$

$e = 0$

The parameters are within the validity range given in *Table D.3 of CIDECT Design Guide. 8* for hollow section joints under fatigue loading:

$0.30 \leq \beta \leq 0.60$

$24.0 \leq 2\gamma \leq 60.0$

$0.25 \leq \tau \leq 1.00$

$30° \leq \theta \leq 60°$

- The modified nominal stress ranges for chord and brace:

$\Delta\sigma_{br,ax} = 1.3 \cdot 16 = 21$ MPa

$\Delta\sigma_{ch,ch} = 1.5 \cdot 55 = 83$ MPa

Step 5: SCF calculation for load condition 1 (basic balanced axial loading)

In the CIDECT Fatigue Design Guide 8, formulae and graphs for different types of joints are given. For this example the geometric stress concentration factors can be calculated as follows:

$$SCF_{ch,ax} = \left(\frac{\gamma}{12}\right)^{0.4} \cdot \left(\frac{\tau}{0.5}\right)^{1.1} \cdot SCF_{0,ch,ax} = 1.16 \cdot SCF_{0,ch,ax}$$

Chord:

where   for $\beta = 0.5$ and $\theta = 30°$: $SCF_{0,ch,ax} = 2.6$

                for $\beta = 0.5$ and $\theta = 45°$: $SCF_{0,ch,ax} = 2.9$

so that   for $\beta = 0.5$ and $\theta = 38.7°$: $SCF_{0,ch,ax} = 2.77$

and $SCF_{ch,ax} = 1.16 \cdot 2.77 = 3.2$

Brace:

$$SCF_{br,ax} = \left(\frac{\gamma}{12}\right)^{0.4} \cdot \left(\frac{\tau}{0.5}\right)^{1.1} \cdot SCF_{0,ch,ax} = 1.08 \cdot SCF_{0,br,ax}$$

3.9 GEOMETRIC STRESS RANGES

where
for $\beta = 0.5$ and $\theta = 30°$: $SCF_{0,br,ax} = 1.3$
for $\beta = 0.5$ and $\theta = 45°$: $SCF_{0,br,ax} = 1.8$

so that for $\beta = 0.5$ and $\theta = 38.7°$: $SCF_{0,br,ax} = 1.59$

and $SCF_{br,ax} = 1.08 \cdot 1.59 = 1.72$

Check minimum SCF value:

for $\theta = 30°$: min $SCF_{br,ax} = 2.64$
for $\theta = 45°$: min $SCF_{br,ax} = 2.30$

so that for $\theta = 38.7°$: min $SCF_{br,ax} = 2.44$

so use minimum SCF value, $SCF_{br,ax} = 2.4$

Step 6: SCF calculation for load condition 2 (chord loading)

From CIDECT Design Guide 8, Table D.3:

Chord:

$$SCF_{ch,ch} = 1.2 \cdot \left(\frac{\gamma}{0.5}\right)^{0.3} \cdot (\sin\theta)^{-0.9} = 1$$

use minimum value, $SCF_{ch,ch} = 2.0$

Brace:

$SCF_{br,ch} = 0$
(negligible)

Step 7: Values of the structural stress ranges at the hot spot for design

For load condition 1 (basic balanced axial loading):

$\Delta\sigma_{hs,ch,ax} = SCF_{ch,ax} \cdot \sigma_{br,ax} = 3.2 \cdot 21$ MPa $= 67$ MPa

$\Delta\sigma_{hs,br,ax} = SCF_{br,ax} \cdot \sigma_{br,ax} = 2.4 \cdot 21$ MPa $= 50$ MPa

For load condition 2: (chord loading):

$\Delta\sigma_{hs,ch,ch} = SCF_{ch,ch} \cdot \sigma_{ch,ch} = 2.0 \cdot 83$ MPa $= 166$ MPa

$\Delta\sigma_{hs,br,ch} = SCF_{br,ch} \cdot \sigma_{ch,ch} = 0$ MPa

Superposition of load conditions 1 and 2:

3. DETERMINATION OF STRESSES AND STRESS RANGES

$$\Delta\sigma_{hs,ch} = 67 \text{ MPa} + 166 \text{ MPa} = 233 \text{ MPa}$$
$$\Delta\sigma_{hs,br} = 50 \text{ MPa} + 0 \text{ MPa} = 50 \text{ MPa}$$

Step 8: Determination of partial resistance factor

A partial factor on stress ranges is required for design. For this example the joint is assumed to be damage tolerant (more than one truss supporting the floor, semi-rigid truss joints, crack sites well defined and regular inspection possible) and with high consequences for a failure (possibilities of partial collapse on people and high financial losses due to shut down of unit). From Table 3.1 of EN 1993-1-9, the partial factor $\gamma_{Mf} = 1.15$ has been taken.

$$\Delta\sigma_{hs,ch} = 1.15 \cdot 233 = 268 \text{ MPa}$$
$$\Delta\sigma_{hs,br} = 1.15 \cdot 50 = 58 \text{ MPa}$$

Step 9: Fatigue life for joint 6

Using the diagram given in CIDECT Fatigue Design Guide 8, page 30 (Figure 3.3), the fatigue life can be determined. From the same publication, the expression for the fatigue strength can be obtained (Table 3.1 in the CIDECT design guide 8) and reads:

$$\log(\Delta\sigma_{hs}) = \frac{1}{3}\left(12.476 - \log(N_R)\right) + 0.06 \cdot \log(N_R) \cdot \log\left(\frac{16}{T}\right) \text{ for } N_R \leq 5 \cdot 10^6$$

Where T is the tube wall thickness

Note: for a tube wall thickness $T = 16$ mm, this expression corresponds to a fatigue strength of 114 N/mm² at 2 million cycles.

For fatigue cracking in the chord, that is with $T = t_0 = 7.1$ mm and $\Delta\sigma_{hs,ch} = 268 \text{ MPa}$, the above expression leads to the following fatigue life:

$$\log(N_R) = \frac{12.476 - 3 \cdot \log(\Delta\sigma_{hs})}{1 - 0.18 \cdot \log(16/T)} = \frac{12.476 - 3 \cdot \log(268)}{1 - 0.18 \cdot \log(16/7.1)} = 5.54$$

Hence, the fatigue life expectancy of joint 6 is $N_R = 10^{5.54} = 346\,000$ cycles, with failure in the chord.

In comparison to the result obtained with the classification method, around 46 000 cycles, it can be seen that effectively the use of the rules in EN 1993-1-9 result in a more conservative estimation of the fatigue life.

Chapter 4

FATIGUE STRENGTH

4.1 INTRODUCTION

4.1.1 Set of fatigue strength curves

It has been seen in chapter 1 that the statistical analysis of the test results on a specific structural detail allowed for the definition of one fatigue strength curve (Figure 1.3). Numerous fatigue tests programs on different details in steel have shown that the fatigue strength curves are more or less parallel. Fatigue strength is thus only a function of the constant C, see equation (4.1), which value is specific to each structural detail.

$$\log N = \log C - m \cdot \log_{10}(\Delta\sigma) \qquad (4.1)$$

Since there are many different details, so is the number of the different strength curves, and this is unusable for design in practice. The solution is the classification of the different structural details in categories with a corresponding set of fatigue strength curves.

Classified structural details may be described in different EN 1993 associated Eurocodes (EN 1993-1-9, EN 1993-2, EN 1993-3-2, etc.) but they all refer to the same set of fatigue strength curves, as given in the generic part 1-9. Each detail category corresponds to one S-N curve where the fatigue strength $\Delta\sigma$ is a function of the number of cycles, N, both represented in logarithmic scale. The set is composed of 14 S-N curves, equally spaced in log scale. The set has been kept the same over the last decades; it comes from the ECCS original work of drafting the first

4. Fatigue Strength

European recommendations (ECCS, 1985). The set is reproduced in Figure 4.1. The spacing between curves corresponds to a difference in stress range of about 12 % (values corresponding to the detail categories were rounded off), i.e. 1/20 of an order of magnitude on the stress range scale.

All curves composing the set are parallel and each curve is characterized, by convention, by the detail category, $\Delta\sigma_C$ (value of the fatigue strength at 2 million cycles, expressed in N/mm^2). It is also characterized by the constant amplitude fatigue limit (CAFL), $\Delta\sigma_D$, at 5 million cycles, which represents about 74 % of $\Delta\sigma_C$. The slope coefficient m is equal to 3 for lives shorter than 5 million cycles. For constant amplitude stress ranges equal to or below the CAFL, the fatigue life is infinite.

The CAFL is fixed at 5 million cycles for all detail categories. This is not exactly the case in real fatigue behaviour but has advantages for damage sum computations. Other codes use different values. For example the AISC code uses values ranging from 1.8 to 22 million cycles depending upon the detail category (the lower the category, the higher the number of cycles for the CAFL) (AISC, 2005).

Under variable amplitude loadings, the CAFL does not exist, but still has an influence. Thus, a change in the slope coefficient is made, the value $m = 5$ being used between 5 million and 100 million of cycles. This last value corresponds to the cut-off limit, $\Delta\sigma_L$, which corresponds to about 40 % of $\Delta\sigma_C$. By definition, all cycles with stress ranges equal to or below $\Delta\sigma_L$ can be neglected when performing a damage sum. The reason for this is that the contribution of these stress ranges to the total damage is considered as being negligible. It should be emphasised that the double slope S-N curve (and the cut-off limit), compared to the unique slope curve, represents better the damaging process due to cycles below the constant amplitude fatigue limit (CAFL). This is in particular valid when the spectra follow a distribution close to Rayleigh's, which is usually the case for civil engineering structures. For other types of spectra such as in aeronautics, with transient load fluctuations during take-off/landing, this representation is not adequate.

It should be again emphasised that the behaviour under variable amplitude loading is complex. A few stress cycles can influence the start of a fatigue crack, even though the contribution of these very same cycles to the damage sum is negligible (see section 5.4.2).

4.1 INTRODUCTION

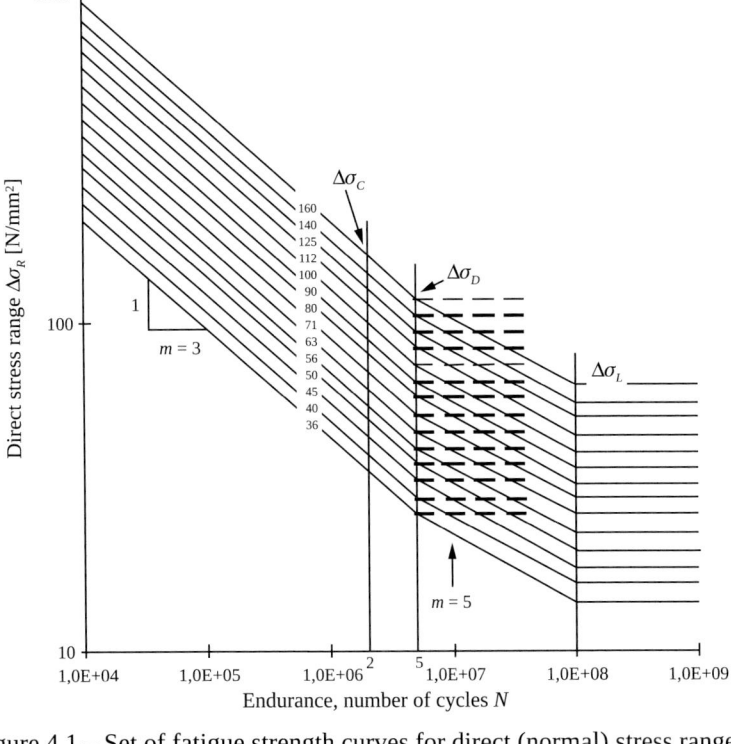

Figure 4.1 – Set of fatigue strength curves for direct (normal) stress ranges

If a structural detail configuration from a type of structure can be found in the tables of the relevant EN 1993 associated Eurocodes, and the description and requirements for this detail correspond, then the fatigue strength can be derived from the standard fatigue resistance S-N curves given in EN 1993, generic part 1-9.

Note that these fatigue curves are based on representative experimental investigations. They include the effects of:

- stress concentrations due to the detail geometry (detail severity),
- local stress concentrations due to the size and shape of weld imperfections within certain limits,
- stress direction,
- expected crack location,
- residual stresses,
- metallurgical conditions,
- welding and post-welding procedures.

4. FATIGUE STRENGTH

Additional stress concentrations due to geometry and not included in the classified structural details, e.g. misalignment, large cut-out in the vicinity of the detail, have to be accounted for by the use of a stress concentration factor. The stress concentration factor is usually put on the action effects side but not always, as explained in sub-chapter 3.4.

There are several specific cases not following the above set of fatigue strength curves. A first case is the tubular lattice girder node joints (EN 1993-1-9, Table 8.7), which is treated separately since the slope coefficient found to represent the fatigue behaviour is $m = 5$. As a consequence, there is no slope coefficient change at 5 million cycles in the S-N curve for these details. The set of S-N curves for these details is given in Figure 4.2 below.

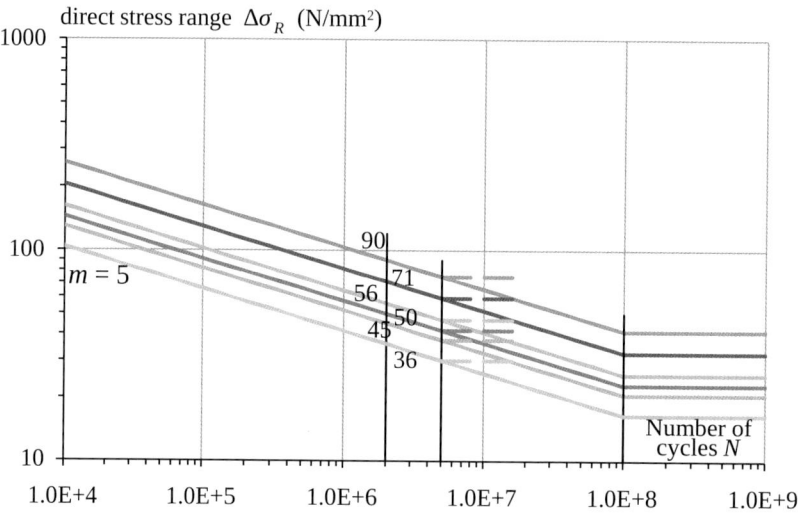

Figure 4.2 – Set of fatigue strength curves for tubular lattice girder node joints (details from Table 8.7)

Another specific case is tension components, mainly cables, for which there is a specific set of fatigue strength curves, see sub-section 4.2.10 for more information.

For shear stress ranges, the statistical analysis of the test results on specific structural details with fatigue cracks developing under shear have shown differences with those under direct or normal stress ranges. Firstly,

4.1 INTRODUCTION

the fatigue strength curves slope coefficient is higher than under direct or normal stress ranges, leading to a slope coefficient $m = 5$. Secondly, there is no well defined constant amplitude fatigue limit and thus the curve has no CAFL. Thirdly, as for the other S-N curves, there is a cut-off limit at 100 million cycles. There are only a few details in shear only (Table 8.1, details 6, 7 and 15, Table 8.5, details 8, 9) so only two fatigue strength curves are needed to classify them as shown in Figure 4.3. However, there is a third, very special, S-N curve for studs in shear (detail 10, Table 8.5), with a slope coefficient $m = 8$, no CAFL and no cut-off limit, also shown in Figure 4.3. A cut-off limit would not change significantly the fatigue verification since the slope coefficient is very high, which explains why it is not specified.

Each curve is characterized by convention, again, by the detail category, $\Delta\tau_C$ (value, expressed in N/mm², of the fatigue strength at 2 million cycles). The curve with a unique slope coefficient, $m = 5$, is used up to 100 million cycles. This number of cycles corresponds to the cut-off limit, $\Delta\tau_L$. This means, again, by definition, that all cycles having stress ranges below $\Delta\tau_L$ can be neglected when performing a damage sum for the same reason as before.

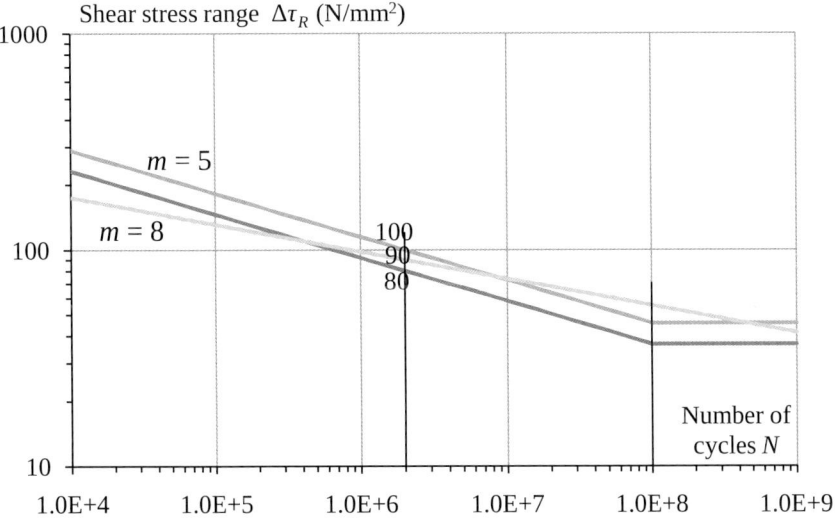

Figure 4.3 – Set of fatigue strength curves for shear stress ranges

4. FATIGUE STRENGTH

4.1.2 Modified fatigue strength curves

The fatigue resistance of a few details however do not fit well in the original set of fatigue strength curves. Thus, modified curves have been added. An example of one of these modified fatigue strength curves, category 45*, is represented in Figure 4.4. The main difference is the location of the CAFL. The detail category is kept the same (at 2 million cycles), so are the slope coefficients ($m = 3$ and 5), but the CAFL as well as the slope change is located at 10 million cycles instead of 5 million. For lives over 10 millions cycles, as said before, the slope coefficient m changes from 3 to 5, until 100 million cycles, where the cut-off limit is reached.

With the rules given in EN 1993-1-9, the following two approaches can be chosen for such details:

- the detail category is put in the original set of curves. It results in a conservative approach when doing the verification with respect to fatigue strength at 2 million cycles. But it will result in a non-conservative verification if the CAFL is used.
- the detail category is put in the upper class, since it has an asterisk, $\Delta\sigma_C^*$, and the CAFL has now to be computed at 10 million cycles. This results in a lower CAFL value compared to the previous approach. The following equivalence can be written:

$$\Delta\sigma_D (\text{at 10 million cycles}) = (2/10)^{1/3} 1.12 \Delta\sigma_C^* \qquad (4.2)$$

In this case, the verification using the CAFL will be correct as well as the verification with respect to fatigue strength at 2 million cycles (and more economical).

One must be careful when using the first approach. As an example, one can look at an overlapped joint (detail 5, Table 8.5), which has a detail category 45*. This means that this detail can be conservatively classified as a category 45 detail. But, alternatively, it can also be classified as a category 50, providing that its CAFL is taken as $(2/10)^{1/3} 50 = 29$ N/mm² at 10 million cycles. Both classification cases are drawn for comparison in Figure 4.4. The values of the conservative and alternative classifications given in EN 1993-1-9 are summarised in Table 4.1.

4.1 INTRODUCTION

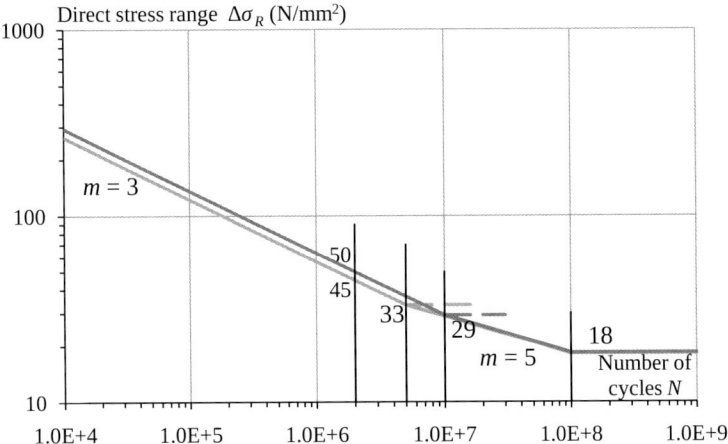

Figure 4.4 – The two alternative fatigue strength curves for a particular detail category 45* under direct (normal) stress ranges

Table 4.1 – Modified strength curves, original and alternatives values

Category	Original curves [N/mm²]			Alternative curves [N/mm²]		
	$\Delta\sigma_C$	$\Delta\sigma_D$ at $5\cdot10^6$	$\Delta\sigma_L$	$\Delta\sigma_C$	$\Delta\sigma_D$ at 10^7	$\Delta\sigma_L$
36*	36	26.5	14.6	40	23.4	14.6
45*	45	33.2	18.2	50	29.2	18.2
56*	56	41.3	22.7	63	36.8	22.7

4.1.3 Size effects on fatigue strength

The influence of the size of the detail on its fatigue strength is recognized in different ways. Firstly, the test results used to fix the fatigue strengths of the details were carried out on specimens with dimensions that are sufficient to represent correctly the built-in welding residual stresses. Secondly, some details in the tables have been separated according to the variation of one or two geometrical dimensions; for example a longitudinal attachment can correspond to four different categories according to the attachment length (see Table 4.2). This can be called a non-proportional scaling effect, since only some dimensions are scaled and not the others.

4. FATIGUE STRENGTH

Table 4.2 – Influence of length on the detail category for a longitudinal attachment (extract from Table 8.4 EN 1993-1-9)

Detail category	Constructional detail		Description
80	$L \leq 5$ m		Longitudinal attachments:
71	$50 < L \leq 80$ mm		1) The detail category varies according to the length of the attachment L.
63	$80 < L \leq 100$ mm		
56	$L > 100$ mm		
71	$L > 100$ mm $\alpha < 45°$		2) Longitudinal attachments to plate or tube.

Thirdly, for cases that are close to proportional scaling, one can see that the size effect in fatigue is essentially influenced by the plate thickness in which the fatigue crack grows and therefore has often been called the "thickness effect". For these cases, the reduction formula for size effects suggested originally by Gurney (1979) is used:

$$\Delta\sigma_{C,red} = k_s \cdot \Delta\sigma_C \tag{4.3}$$

with

$$k_s = \left(\frac{t_0}{t}\right)^n < 1.0 \tag{4.4}$$

The thickness t_0 is the reference thickness above which a reduction due to size effect has to be considered; it is usually taken as 25 mm. The value of the exponent n in the formula (4.4) is function of the detail considered.. It can take values comprised between 0.1 and 0.4 depending upon the detail considered (the exponent increases proportionally to the stress concentration factor at the crack location) (IIW, 2009). In EN 1993-1-9, it is equal to 0.2 for butt joints and 0.25 for bolts in tension (but in this case with a reference diameter of 30 mm, see section 4.1.3 for further information).

The size effect reduction can be demonstrated by taking as an example a transverse butt weld, or splice, joining two plates of different thicknesses, namely 100 and 60 mm. With proper requirements, this detail can be taken as detail 5 from Table 8.3 of the EN 1993-1-9 (see Annex B.3), and thus classified in category 90. The correct, reduced, fatigue strength due to its size is found by applying formula (4.4) with $t = 60$ mm as it is the side where the fatigue crack will develop, which gives $(25/60)^{0.2}\ 90 = 75.5$ N/mm^2.

For geometric stress approach, EN 1993-1-9 does not mention the effect of component size. However, depending upon the extrapolation method and the type of joints, a geometrical size effect should be taken into account. For hot-spot stress type a, the multiplying factor is identical to the one given in formula (4.4). Indications are given in Annex B, Table B13. Note that the extrapolation method using fixed points is also intended to take into account the geometric size effect) (Niemi *et al*, 2006). For hot-spot stress type b, the plate thickness has only a small effect on fatigue strength, because the geometrical effect now depends mainly on the width of the plate.

4.1.4 Mean stress influence

The mean stress influence is only relevant in the cases of non-welded details or details on which post-weld improvements have been made. It can be accounted for by acting on the action effects side or on the fatigue strength side. For the case of non-welded details, the modification of the stress range is given in section 3.7.2, that is the modification is made on the action effects side. The case of post-weld improvement is explained in the next section.

4.1.5 Post-weld improvements

Where the classification in tables 8.1 to 8.10 of EN 1993-1-9 does not give adequate fatigue strength, the performance of weld details may be improved by post-welding treatments such as controlled machining, grinding or peening. When this is required, and when the proposed improvement method is not covered by EN 1993-1-9, the detail should be classified by testing (see sub-chapter 4.3).

A family of post-weld improvement techniques such as needle peening, shot peening or UIT (ultrasonic impact treatment), introduce compressive residual stresses in the surface layer where the fatigue cracks starts. Thus, the fatigue strength of these improved details is influenced by the mean stress of the applied external loads and has to be properly accounted for. It is outside of the scope of this book to present post-weld improvement methods. Thus, for further information on recommended procedures, quality control, fatigue

4. FATIGUE STRENGTH

strength categories, etc. the reader is advised to read the relevant IIW recommendations (IIW, 2009.b). Note that IIW proposes a modification in the definition of the stress range together with a change of classification of the improved detail. Thus, the mean stress effect is accounted for by acting on both sides of the verification relationship.

The authors would emphasise here that improvement techniques should be thought of right at the initial design stage, especially to compensate for bad detailing or fabrication, but only once other possibilities have been unsuccessfully explored. These methods are for example very useful when designing structures with high strength steels. But they can also represent a useful option when the need for an increase in fatigue life is discovered, for example at a late stage of fabrication, when the structure is already in service, as a fatigue retrofit or strengthening option (after evaluation, NDT controls and under given conditions only).

4.2 FATIGUE DETAIL TABLES

4.2.1 Introduction

This sub-chapter gives useful information on the detail classifications given in tables 8.1 to 8.10, including notes on the potential modes of failure, important factors influencing the class of each detail type and some guidance on selection for design. It is a synthesis of information taken from the following sources: BS 7608, NORSOK 004 and draft of the background document to EN 1993-1-9, BS 7608 (1993), NORSOK (2004) and Stötzel *et al* (2007). Commentary specific to each detail, that is clarification and advice for performing correctly the verification, have been included directly in the detail category tables given in annex B of this book.

4.2.2 Non-welded details classification (EN 1993-1-9, Table 8.1)

In unwelded details, fatigue cracks normally initiate either at surface irregularities, at corners of the cross sections, or at holes and re-entrant corners. In members connected with rivets or bolts loaded in shear, failure generally initiates at the edge of the hole and propagates across the net section, see Figure 4.5. Fatigue crack may also initiate in the bolt itself.

However, in double covered joints made with high strength friction grip bolts these modes of failure are eliminated by the pre-tensioning (providing joint slip is avoided) and failure may initiate on the surface near the boundary of the compression ring due to fretting under repeated strain, see Figure 4.5. In these details, fatigue crack may also initiate at a geometrical change near or at the bolted coverplates ends.

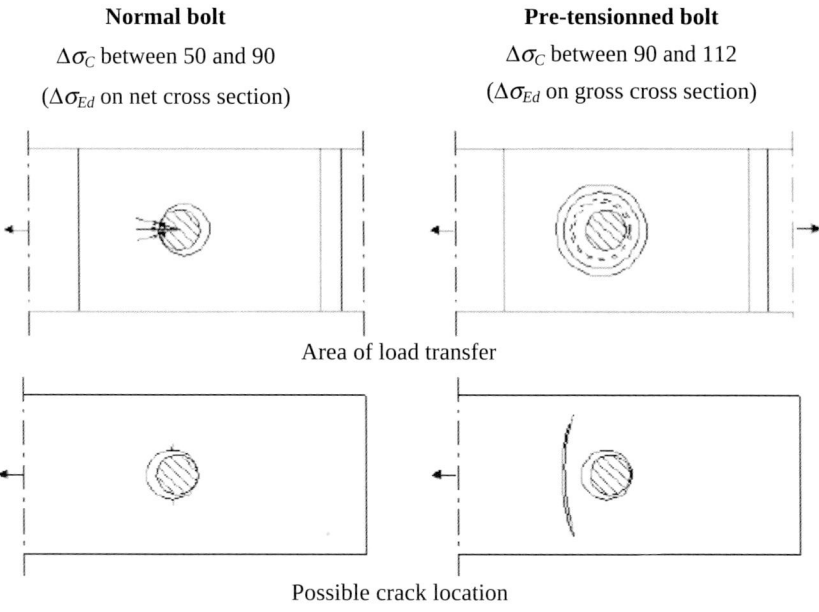

Figure 4.5 – Bolted connections in shear, categories, area of load transfer and possible location of fatigue crack (from ESDEP courses (ESDEP, 1995))

In threaded fasteners, loaded in tension (or combination of tension and shear, with or without secondary bending), fatigue cracks normally initiate at the root of the thread, particularly at the first load carrying thread in the joint. Alternatively, failure is sometimes located immediately under the head of the bolt, particularly in bolts with rolled threads and in joints with bolts subjected to prying effects (secondary bending). It is important to ensure that the specified fit-up of bolted connections is achieved in practice. Otherwise the stress ranges applied to the bolts may be much higher than those assumed in design and hence lead to premature failure. Figure 4.6 shows the different possible fit-up cases and their influence on the fatigue strength of the connection.

4. FATIGUE STRENGTH

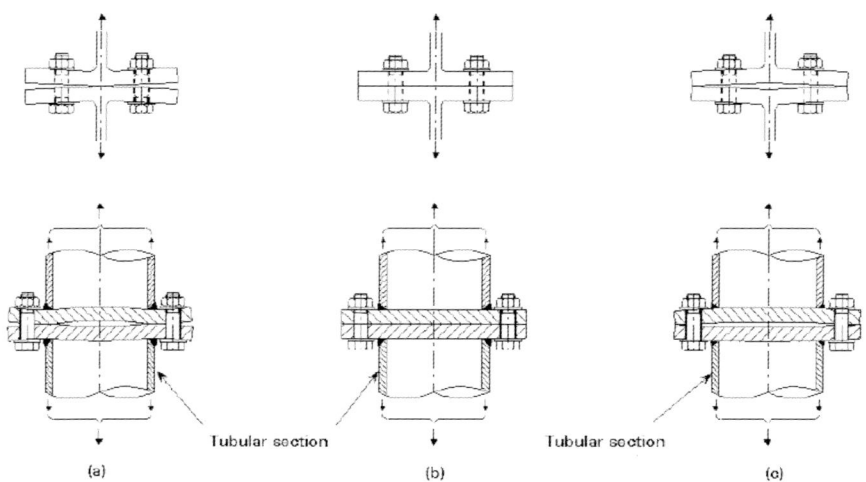

Figure 4.6 – Different possible fit-up cases in bolted connections in tension with (a) best, (c) worst, because of prying forces (from ESDEP courses (ESDEP, 1995))

There is a significant influence of size for bolts and rods in tension. This is due to incomplete geometric scaling of the micro-geometry of the thread but – the thread or notch radius being scaled up to the thread pitch, rather to the diameter for standard (ISO) bolts – the local stress at the notch is a function of the diameter to notch radius. The resulting effect is a decrease of fatigue strength with increasing diameter, expressed similarly to plate thickness influence (see section 4.1.3) as follows:

$$\frac{\Delta\sigma_{C,red}}{\Delta\sigma_C} = \left(\frac{\varnothing_0}{\varnothing}\right)^n \quad (4.5)$$

where

n scale effect exponent, function of the stress concentration, taken as 0.25 (in reality ranging from 0.1 to 0.33, depending upon different parameters such as the type of threads, cut or rolled, the material, etc.)

$\varnothing_0$ reference diameter of bolt, taken as 30 mm in EN 1993-1-9

4.2.3 Welded plated details classification (general comments)

In welded construction, fatigue failure will rarely occurs in a region of unwelded material since, as can be inferred from the previous section 4.2.2,

4.2 FATIGUE DETAIL TABLES

the fatigue strength of the welded joints will usually be much lower because of the presence of discontinuities. The fatigue strength, or joint classification, is directly linked with stress concentration. The welded details with the lowest stress concentration are full penetration transverse butt welds (first details in Table 8.3). Particularly high increase in stress concentration, hence reduction in fatigue strength, occur where the following features are present:

- the weld ends or toes are on, or near, an unwelded corner of the member. This is the reason why an 'edge distance' is specified for some of the joints;

the attachment is 'long' in the direction of the direct stress and, as a result, transfer of a part of the load in the member to and from the attachment will occur through welds adjacent to its ends, see Figure 4.7. Parallel fillet welds have better fatigue strength (along attached plate in case b)), followed by long transverse welds (case d) and if full penetration case e)), and finally the worst are the short transverse welds (ends of longitudinal attachments, case a), end transverse weld in case b), and case c)).

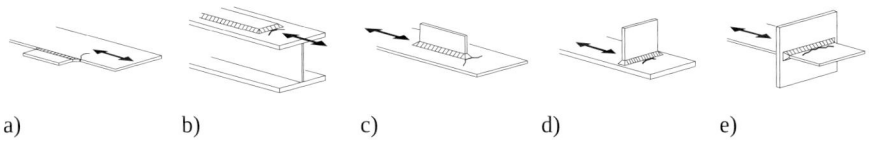

a)   b)   c)   d)   e)

Figure 4.7 – Different types of welded details

Furthermore, the welded plated details are classified in the different tables according to the following:

- Welds between plates in the same plane (transverse butt joints)
- Welds not in the same plane as the plates, that is T-joints or lap joints.
- Welded attachments loaded or not.

The following paragraphs present the regrouping of the details contained in the different classification tables, with the numbering from the code.

4.2.4 Longitudinal welds, (built-up sections, EN1993-1-9 Table 8.2), including longitudinal butt welds

Regarding the potential modes of failure, away from weld ends, fatigue cracks normally initiate at stop-start positions or, if these are not present, at weld surface ripples. With the weld reinforcement dressed flush, failure tends to be associated with weld flaws. However, in the case of discontinuous welds (details 8 and 9, Table 8.2) fatigue cracks will occur at the weld ends.

No edge distance criterion exists for continuous or regularly intermittent welds away from the ends of an attachment. However it is important to limit the possibility of local stress concentrations occurring at unwelded corners as a result of, for example, undercut, weld spatter and excessive leg length at stop-start positions or accidental overweave in manual welding.

Although this criterion can be specified only for the 'width' direction of a member, it is equally important to avoid undercutting on the unwelded corners of, for example, cover plates wider than the flange on which they are welded. If it does occur, it should subsequently be ground out to a smooth profile.

4.2.5 Transverse butt welds (EN1993-1-9 Table 8.3)

With the ends of butt welds machined flush to the plate edges (after removal of weld run-on and run-off pieces), fatigue cracks normally initiate at the weld toe. They then propagate into the parent metal, so that the fatigue strength depends largely upon the toe profile of the weld. If the reinforcement of a butt weld is dressed flush, failure is more likely to occur in the weld material from embedded flaws or from minor weld flaws which become exposed on the surface, e.g. surface porosity in the dressing area (typical for details 1 to 3 of Table 8.3). In the case of butt welds made on a permanent backing strip, fatigue cracks initiate at the weld metal strip junction (weld root) and then propagate into the weld metal (details 14 to 16).

The classification, as given explicitly in BS 7608 (1993), may be deemed to allow for the effects of any accidental axial or centreline misalignment up to the lesser of 0.15 times the thickness of the thinner part or 3 mm, provided that the root sides of joints with single-sided preparations i.e. single bevel, –J, –U or –V, are back-gouged to a total width at least equal to half the thickness of the thinner member.

However, where such support is not provided, e.g. tension links, and where the amount of misalignment exceeds the limits stated above, the design stress should include an allowance for the bending effects of any intentional misalignment, i.e. the nominal distance between the centres of thickness of the two abutting members. For members tapered in thickness, the mid-plane of the untapered section should be used. The nominal stress should be multiplied by the factor k_s, as explained in section 3.7.7 or given for detail 17.

For other cases, including angular misalignment, see sub-chapter 3.4.

4.2.6 Welded attachments and stiffeners (EN 1993-1-9 Table 8.4), and load-carrying welded joints (EN 1993-1-9 Table 8.5)

For fillet welds, it can be seen in the classification that the weld direction, parallel to the main stress flow or perpendicular to it, and its length in the perpendicular direction influence significantly the detail class.

For transverse joints, the overall joint width should be minimised as much as possible, for example, by using partial penetration welds instead of fillet welds when multi-pass welds are needed (refer to Figure 4.8). The loaded plate can also be interrupted as also shown in the figure.

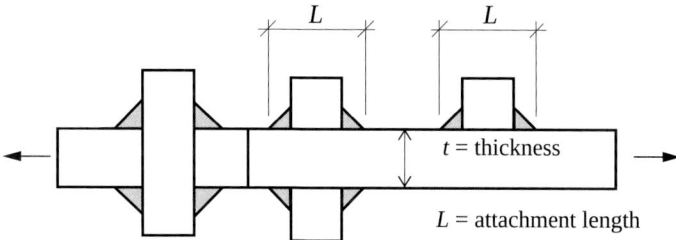

Figure 4.8 – Total attachment length, L, in transverse attachments

For longitudinal attachments, significant improvement in the fatigue strength can be achieved by shaping the ends of the gusset and also by grinding properly the weld toe as well. Note that the improvement is for the toe only, and for large increase in fatigue strength of the detail, the possibility of fatigue cracking from the weld root shall also be addressed (i.e. will limit the fatigue strength increase).

In the case of welded shear connectors, fatigue cracking tends to occur either in the weld throat initiating from the root due to shear and associated bending transmitted from the slab (detail 10, Table 8.5) or at the weld toe and propagating in the flange for highly stressed flanges (detail 9, Table 8.4). The combined effects of the two possible fatigue cracking modes is accounted for by using the interaction formulas for verification under multiaxial stress ranges of welded studs presented in sub-section 5.4.7.4.

4. Fatigue Strength

Example 4.1: Detail classification and verification, application to chimney (socket joint located at +11490 mm) (worked example 2)

Fatigue strength of the details

A detailed description of the different fatigue details composing the socket joint was made in chapter 1. The socket joint with fillet welds corresponds to detail number 12, Table 8.5, see Figure 4.9 below. This is a detail category 40. Thus:

$$\Delta\sigma_C = 40\,\text{N/mm}^2 \text{ and CAFL}: \Delta\sigma_D = 0.74 \cdot \Delta\sigma_C = 29.6\,\text{N/mm}^2$$

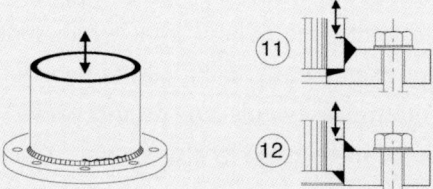

Figure 4.9 – Chimney socket joint fatigue detail (EN 1993-1-9, Table 8.5)

Bolt in tension corresponds to detail number 14, Table 8.1, see below. This is a detail category 50 up to a bolt diameter of 30 mm. Thus: $\Delta\sigma_C = 50\,\text{N/mm}^2$

CAFL for this detail is $\Delta\sigma_D = 0.74 \cdot \Delta\sigma_C = 37\,\text{N/mm}^2$

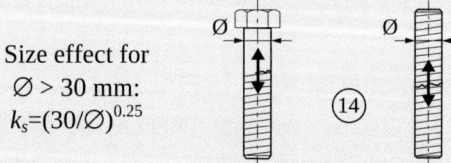

Size effect for
Ø > 30 mm:
$k_s = (30/\text{Ø})^{0.25}$

Figure 4.10 – Fatigue detail for bolt in tension in chimney socket joint (EN 1993-1-9, Table 8.1)

The connection is checked against fatigue in example 5.1.

Example 4.2: Detail classification, application to chimney (bottom socket joint located at +350 mm) (worked example 2)

In this example, it will be shown how to classify the different details composing a complex welded connection. The geometry and dimensions of the chimney are given in section 1.4.3 and, for the bottom socket joint, in sub-section 1.4.3.4.

4.2 FATIGUE DETAIL TABLES

The wind loads are computed in section 3.1.5 (example 3.1), the number of cycles in example 3.5. The connection is checked against fatigue in example 5.2.

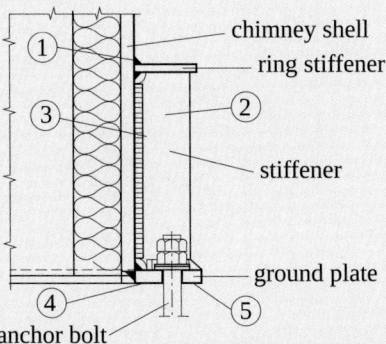

Figure 4.11 – Chimney bottom socket joint, with possible fatigue crack locations

1) Upper ring welded to shaft. Transverse attachment, classify as detail 6, Table 8.4, category 80 ($\ell < 50$ mm). Note that since longitudinal stiffener continues below the upper ring, if there would be no mouse-hole, this detail would classify better as a longitudinal attachment than a transverse one.
2) Mouse-hole in stiffener in order to properly weld the ring and ground plate. Detail 9, Table 8.2, category 71.
3) Longitudinal fillet weld along stiffener. Detail 7, Table 8.2, category 100 (this detail is never the critical one).
4) Welded detail between longitudinal stiffener and ground plate. The stress flow being in the stiffener, this detail is a Tee-butt joint with fillet welds and $\ell = 40 + 8.5 < 50$ mm, detail 3, Table 8.5. Corresponding details:
   - root cracking: category 36* and 80
   - toe cracking: category 80.
5) Anchor bolt in tension with, may be some bending due to prying effects. Detail 14, Table 8.1, category 50.

Example 4.3: Application to runway beam of crane (worked example 3)

At this stage, the construction details should be identified having in mind that the welding and stress concentrators can reduce considerably the fatigue life of the runway steel beam.

4. FATIGUE STRENGTH

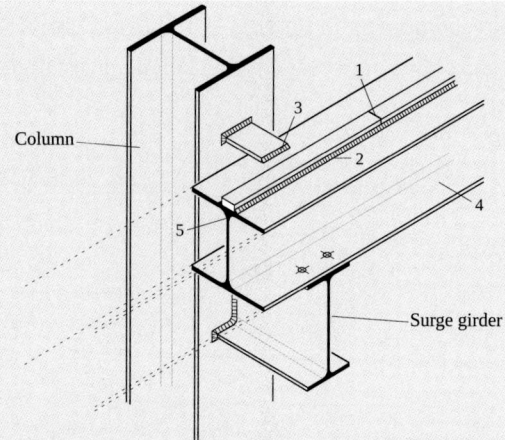

Figure 4.12 – Geometry of runway beam of crane

Classification of the detail categories

The fatigue critical details are identified as:

1. Transverse joint of the rail

$\Delta\sigma_C = 71\,\text{N}/\text{mm}^2$ (detail 14 from Table 8.3 EN 1993-1-9)

2. Continuous longitudinal weld between the rail and the top flange

$\Delta\sigma_C = 125\,\text{N}/\text{mm}^2$ (detail 2 from Table 8.2 EN 1993-1-9)

3. Transverse support of the top flange

$\Delta\sigma_C = 40\,\text{N}/\text{mm}^2$ (detail 5 from Table 8.4, EN 1993-1-9)

4. Runway rolled beam, flange due to bending moment

$\Delta\sigma_C = 160\,\text{N}/\text{mm}^2$ (detail 2 from Table 8.1, EN 1993-1-9)

5. Beam web under local vertical stress $\sigma_{z,local}$

$\Delta\sigma_C = 160\,\text{N}/\text{mm}^2$ (detail 1 from Table 8.10, EN 1993-1-9)

4.2.7 Welded tubular details classification (EN 1993-1-9 Tables 8.6 and 8.7)

For welded tubular connections, the very limited geometrical validity ranges from Tables 8.6 and 8.7 (in particular tube thicknesses up

4.2 FATIGUE DETAIL TABLES

to 8 mm only) limit their practical use to small structures. Furthermore, the nominal stress approach does not properly account for the complex stress field present in tubular joints, that is the many different stress concentrations and stress gradients which are further function of the joint loading conditions. Thus, for other applications such as wind towers, bridges, etc. the geometric stress (structural stress at the hot spot) approach is more appropriate, see sub-chapters 3.5 and 3.9. Since EN 1993-1-9 does not contain in its annex B specific information for tubular joints, one shall seek this information in the CIDECT recommendations (CIDECT, 2001), IIW recommendations (IIW, 2000) and other published literature (FOSTA, 2010).

4.2.8 Orthotropic deck details classification (EN 1993-1-9 Tables 8.8 and 8.9)

For orthotropic deck connections, Table 8.8 gives the detail categories for closed trough and Table 8.9 gives the detail categories for open trough (or stringers). Orthotropic decks contain different fatigue details and a distinction can be made between cracks that are caused in a load-carrying member or in a connection for load transfer, and cracks that are generated by imposed deformations. The different possible cracks locations are given in EN 1993-2 and summarised in Figure 4.13. For the determination of the stresses relevant to the detail under study, one can use a nominal or a geometric stress approach, as long as it is coherent with the detail classification chosen from EN 1993-1-9. For the application of the geometric stress method, the detail category has to be chosen with regard to the different possible weld types as explained in section 4.2.11.

However, since the information contained in EN 1993-1-9 and EN 1993-2 is difficult to grasp, revised tables have been developed by the authors to summarize the information and to guide the engineer, see Annex B, Table B.13. They include notes on the potential modes of failure (crack location and consequences), important factors influencing the class of each detail type and some guidance on selection for design, including strength factor γ_{Mf}. These tables are a combination and interpretation of the standards as well as propositions from different recent studies (Kolstein, 2007) and (Leendertz, 2008).

4. FATIGUE STRENGTH

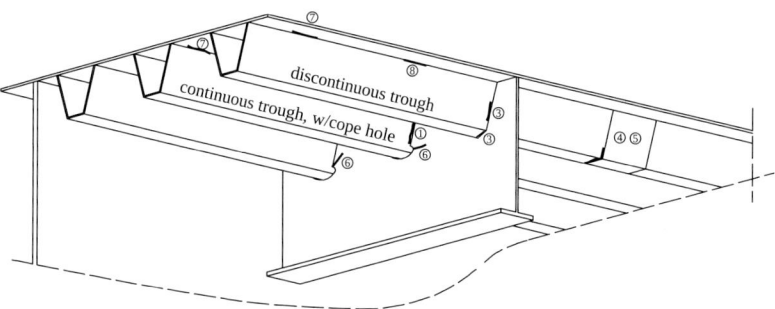

Figure 4.13 – Typical orthotropic steel deck crack locations, adapted from EN 1993 – 2 (detail numbering corresponds to the one in Table B.13)

4.2.9 Crane girder details (EN 1993-1-9 Table 8.10)

This table contains the detail categories of the flange to web junction of runway beams subjected to vertical compressive stress range due to wheel load (local effect). Fatigue cracking is shown to always start from the detail weld toe in the figures, but could also start from an internal flaw or non-welded zone in the case of details with fillet welds. Accordingly, the stress range is to be computed either in the web or in the weld throat. The fatigue crack propagates horizontally, along or in the weld, and is assumed not to be influenced by the beam bending stresses. Fatigue verification is thus performed separately from those other fatigue cracking cases, see sub-section 5.4.7.3.

4.2.10 Tension components details (EN 1993-1-11)

This is a special case for which there is a specific set of fatigue strength curves, different from welded joints. Cables in particular are structural elements that can be subjected to two different types of fatigue rupture that follow different laws and have to be considered separately (Cluni *et al*, 2007):

- axial fatigue and bending fatigue. Axial fatigue is originated by fluctuations in the axial tension.
- bending fatigue is the consequence of the combination of axial preload and cyclic bending occurring near the anchors, where the cables behave as clamped; it is investigated by the rotative bending test.

In both axial and bending fatigue, fatigue phenomenon is seen to occur by fretting, thus indicating that the cracks are mainly caused by the friction stresses originated by the sliding contact between wires. Another aspect that distinguishes cables from welded joints is the fact that the former do not present fragile failures since the rupture is preceded by numerous wire failures.

In part 1-11 of Eurocode 3, detail categories are given for different tension components, however not all; they are listed in Table 4.3. Figures of the tension components can be found in Annex C of part 1-11. A summary has been made in the form of a table in Annex B of this manual, Table B.12. Even if it does not stand out from the table, strands, made of parallel wires (group C), have a better intrinsic fatigue resistance than ropes, made of wires in spiral configuration (group B). It is because spiral configuration induces relative displacements between wires even under pure tension and thus more interwire fretting fatigue.

Table 4.3 – Groups of tension components and corresponding detail categories for fatigue strength

Group	Main tension element	Component	Detail category [N/mm^2] for exposure class 3 or 4
A	Single solid round cross section connected to end terminations by threads	Tension rod (bar) system	***
		Prestressing bar	105
B	Ropes composed of wires or stands, in spiral and which are anchored in sockets or other end terminations	Spiral, circular strand rope*	***
		Fully locked circular coil rope with metal or resin socketing**	150
		Strands with metal or resin socketing	150
C	products composed of parallel wires or parallel strands needing individual or collective anchoring and appropriate protection	Parallel wire strand (PWS) with epoxy socketing	160
		Bundle of parallel strands (seven wire prestressing)	160
		bundle of parallel wires	160
		Multiple bars	***

*	typical diameter range: 5 mm to 160 mm
**	typical diameter range: 20 mm to 180 mm
***	to be determined by tests. Specific requirements for fatigue testing of wire, strands, bars and complete tension components are given in EN 1993-1-11, Annex A.

Fatigue failure of cable systems usually occurs at anchorages, saddles or clamps. The basic requirement is that the fatigue resistance of terminations and anchorages exceeds that of the components. The effective category of detail at these locations should preferably be determined from tests representing the

4. Fatigue Strength

actual configuration used and reproducing any flexural effect or transverse stresses likely to occur in practice, see also sub-chapter 4.3. In the absence of tests, the detail categories from, which conform to the fatigue strength curve family given in Figure 4.14, may be used. It should be emphasized that for elements made of high-strength steel such as wires and ropes a discussion on the existence of a fatigue limit is still open. In some cases, moreover, such as in the case of fretting corrosion, there is no endurance limit, thus highlighting the need for a good protection of the cable from corrosion phenomena (Cluni et al, 2007). As a result, the fatigue curves for cables do not have any CAFL but are bi-linear even under constant amplitude loadings. Conservatively, under variable amplitude loadings, the same curves are used. Finally, a cut-off limit may be introduced at 100 millions in order to allow for a infinite life design approach.

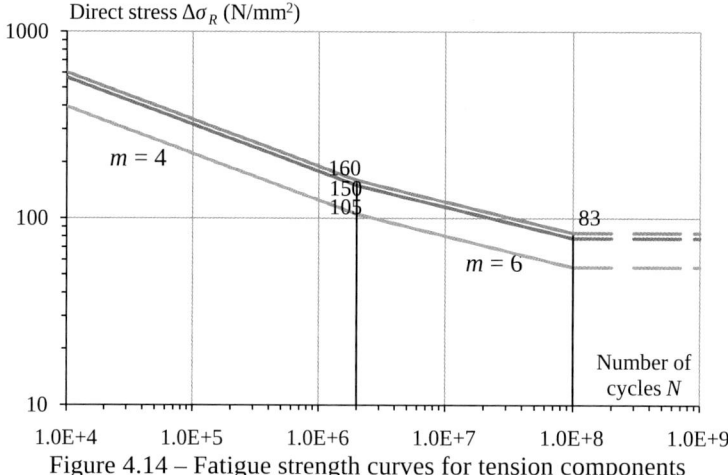

Figure 4.14 – Fatigue strength curves for tension components

The categories given in Table 4.3 are only valid when the following conditions apply:

- design of cables, saddles, cables with sockets and clamps comply with the requirements in EN 1993-1-11;
- large aerodynamic oscillations of cables are prevented using adequate measures such as modification of cable surface, damping devices, stabilizing cables;
- adequate protection against corrosion is provided, see EN 1993-1-11.

For, exposure class 5 according to Table 3.7, that is components subjected to axial and lateral fatigue actions, additional protective measures are required in order to minimize bending stresses.

4.2.11 Geometric stress categories (EN 1993-1-9, Annex B, Table B.1)

Together with the application of the geometric stress method, different possible detail categories exist. The details listed in this table are only to be used for verification if the stresses and stress ranges are determined using the geometric stress approach, as described in sub-chapters 3.5 and 3.9. The category is function of the location of the crack and the geometry of the weld only (e.g. the geometry of the connection being already included in the geometric stress). This results in the following different categories for cracks initiating at:

 – toes of butt welds,
 – toes of fillet welded attachments,
 – toes of fillet welds in cruciform joints.

The detail category table from EN 1993-1-9 is reproduced at the end of this book, Annex B, Table B.11. Note that these classifications are generic and can be used for both plated and tubular joints. However, for tubular joints, it is better to use the CIDECT recommendations (CIDECT, 2001). The CIDECT recommendations make the geometric stress resistance curves dependent upon the tube wall thickness, which is not the case in EN 1993-1-9. However, depending upon the extrapolation method and the type of joints, a geometrical size effect should be taken into account as mentioned in section 4.1.3. For hot-spot stress type a, the multiplying factor is identical to the one given in formula (4.4). In Annex B, Table B.11, the authors have added the most likely thickness correction that should be accounted for, based on the IIW recommendations (IIW, 2009).

The detail categories refer to the as-welded condition, except for detail number 1. The overall weld shape should be similar to the drawings since only the weld configuration differentiates one detail from another in this method, all stress concentrations due to structural imperfections are already included in the geometric stress determination.

In the details given, it is assumed that high tensile residual stresses are present. The cracks are always assumed to start from the weld toe. Cases of cracking from the weld root and propagating through the throat are not covered. Proper detailing and execution must be made to exclude these fatigue cracking cases, for example by following the CIDECT guide recommendations (CIDECT, 2001). Only the effects of small misalignments, up to 5 % stress increase (IIW, 2009), are included in the detail categories

4. FATIGUE STRENGTH

given in EN 1993-1-9 as well as in IIW (same categories). In other words, if the value of the stress concentration factor k_f is less than or equal to 1.05, then it can be neglected. All other effects have to be considered explicitly in the determination of the geometric stresses and stress ranges.

4.2.12 Particular case of web breathing, plate slenderness limitations

Another fatigue problem is the design against cyclic out-of-plane displacements that can occur in slender webs of plate girders under fatigue loads. EN 1993-2 contains a verification formula with a limit on the combination of normal and shear stress ranges values. This fatigue verification is rather complicated and not recommended by the authors but an alternative is proposed. Future traffic load increase, etc. may result in fatigue problems and thus design using EN 1993-2 formula is in the end uneconomical and reduces durability. The alternative consists in plate slenderness limitations in order to avoid any fatigue problems. In order not to have to verify web breathing, the criteria given in equations (4.6) and (4.7) for slenderness in length direction of non-stiffened plates are to be met.

$$b/t \leq 30 + 4.0 \cdot L \text{ and } b/t \leq 300 \text{ for road bridges} \quad (4.6)$$

$$b/t \leq 55 + 3.3 \cdot L \text{ and } b/t \leq 250 \text{ for railway bridges} \quad (4.7)$$

where
- L bridge span in [m] and $L \geq 20$ m,
- b, t plate width and thickness.

The background for these slenderness limitation formulas comes from numerous simulations of damage accumulation made by (Kuhlmann and Günther, 2002) on web plates with imperfections from bridge main girders under realistic load models.

4.3 DETERMINATION OF FATIGUE STRENGTH OR LIFE BY TESTING

Test program objective is either to determinate the fatigue strength or the fatigue life of a constructional detail. General rules about design assisted by testing are given in EN 1990, clause 5.2 and Annex D. However, this Annex D gives only general guidance on test planning, derivation of design values, statistical analysis with influence of number of tests and is more oriented towards static tests. Thus, additional guidance is needed for fatigue testing and analysis; for example, for proper statistical analysis of fatigue data in function of the test program and number of specimens, guidance can be found in the IIW recommendations (IIW, 2009, ISO 12107, 2003).

However, for the specific fatigue testing case of tension components, guidance on fatigue testing can be found in EN 1993-1-11, Annex A. Indeed, the effective detail category of terminations and anchorages should preferably be determined from tests. This is because of the many different detailing of cable anchorage, etc.

As a rule, fatigue testing is long and costly, thus different experimental program designs can be imagined to reduce those. As a result, different statistical analyses must be carried out for experiments with multiple fatigue tests of the same specimen with one detail versus experiments with one fatigue test on a unique beam with multiple identical details, which can be furthermore carried out until failure of all details (with some repairs) or stopped after failure of the first, or second, etc. detail. Furthermore, in order to have realistic residual stress fields and a lower bound for the fatigue strength, it is important to mention that the only proper way to carry out fatigue tests is to perform large or full-scale tests, or to consider properly the size effects (proportional size effects, thickness effect as well as other non-proportional size effects), even if it further increases testing time and costs.

As said before, the first fatigue experimental program purpose is to determine the fatigue strength of a detail or a component. Design assisted by testing is called for studying specific details, fabrication processes, the influence of temperature, loading rate, or their combined effects, etc. In the cases of specific details and fabrication processes, constant amplitude fatigue tests only are to be carried out. One shall now differentiate between fatigue tests to get:

- S-N curve position, and slope, limited life fatigue tests, usually tests are carried out up to 10^6 cycles, or when failure occurs,
- CAFL, fatigue limit, usually tests up to 10^7 cycles,
- High-cycle behaviour, tests must be carried out over 10^7 cycles.

The special case of low-cycle fatigue testing, such as testing to validate a structural detail under earthquake type loadings, is not dealt with here.

4. Fatigue Strength

The scatter in fatigue test results is larger than in static testing. It is usual to find differences of 3 to 5 times in fatigue lives for the same stress range. Thus, at least 10 tests are usually needed to define an S-N curve for design. In order to reduce scatter, specimen fabrication, testing frame and procedures must be well engineered. All the same, an unique failure criteria, such as a fatigue crack length, stiffness change or maximum deflection, must be defined before testing. Note that the statistical analysis may depend on the life domain tested and that proper analysis of fatigue data containing run-outs (tests without failure) is more demanding (ISO, 2003).

In the case of studying the influence of temperature, loading rate, or their combined effects, etc. one may need to carry out variable amplitude tests. All parameters of the stress history may influence the fatigue behaviour and careful determination of the actions and actions effects should be done before any fatigue testing.

The second purpose of fatigue testing is to validate the fatigue design of a component. Again, one can determine fatigue strength, but can also carry out tests to determinate the component fatigue life under realistic loading. In these cases, the variable amplitude test history must be simplified and accelerated (by increasing testing frequency).

This could be the case for tension components where sometimes components and anchorages are especially made for one application. Thus, in EN 1993-1-11, there is only an incomplete table with detail categories and fatigue testing of those components is often required (see Table 4.3). During fatigue tests, no failure should occur in the anchorage material or in any component of the anchorage. Fatigue tests are carried out under tension for exposition classes 3 and 4, and under tension and bending for exposition class 5 (see Table 3.7). In all tests, the maximum stress shall be taken equal to the stress limit given in Table 5.5. A supplementary safety margin is introduced by the requirement of testing cables at $1.25 \cdot \Delta\sigma_C$ and still insure the cable sustains 2 millions cycles (with a number of broken wires less than 2 % of the total). The main size effect which exists is the cable diameter, or the number of wires. Fatigue strength has been shown to decrease with increasing cable diameter (Takena et al, 1992). To avoid another size effect, a statistical one, the test cables need to be long enough, typically a minimum of a couple of meters (Castillo et al, 2006) to include a representative population of imperfections.

Chapter 5

RELIABILITY AND VERIFICATION

5.1 GENERALITIES

The reliability of a structure designed for fatigue decreases with time in service because of the ongoing damage of the structures subject to variable, repetitive loadings.

In general there are different strategies that can be adopted to deal with fatigue reliability. The recommendations published by the International Institute of Welding, for example, propose the following strategies (IIW, 2009):

- *Infinite life design*, based on keeping actions under an assumed fatigue strength limit (i.e. a threshold value). Usually the CAFL, with a partial factor, is used. Most suited for members that experience very high number of cycles, which are preferably close to constant amplitude. A good example is a chimney under vortex-induced vibrations, such as presented in worked example 2. For the structure in service, no regular inspection is theoretically required but may still be specified;
- *Safe life design*, based on the design assumption that no fatigue cracks will form (i.e. cracks stay in the initiation stage) during the whole service life. Used in situations where regular inspection in service is not possible or consequence of failure is very high;
- *Fail safe design*, based on assumption that the structure can tolerate extensive fatigue cracking without failing, possibly because the structure is statically undeterminate or adequately redundant. For the structure in service, regular inspection is needed to detect a failing member before it impairs the structure's serviceability and safety;
- *Damage tolerant design* based on the assumption that fatigue cracks will form but will be readily detectable in service before becoming critical.

5. Reliability and Verification

Thus, the critical crack size (i.e. the material toughness) must be above the detection threshold of the non-destructive testing method applied. Depending upon the structure type, importance, loading, etc. suitable inspection time intervals are prescribed. Once a fatigue crack is detected, it can either be monitored or repaired in function of its criticality.

In EN 1993-1-9, the following two strategies for the verification of fatigue resistance of members, connections and joints subjected to fatigue loading are given:

1) *Safe life method*, providing an acceptable reliability for the structure's design life without the need for regular inspection nor maintenance. In this case, the initially high reliability index level is decreasing with time to reach the minimum (target) value at the end of the design service life;
2) *Damage tolerant method*, ensuring an acceptable reliability that the structure will perform satisfactorily during its design life, provided that inspection and maintenance measures are implemented throughout the life of the structure. In this case, the reliability level – initially lower than that of the *Safe life method* – reaches the minimum target value at the end of the design service life, but with periodical readjustments (using Baysian theory) according to inspection results and possible resulting interventions (repairs).

Figure 5.1 compares the evolution of the reliability index of the safe life and damage tolerant methods over the service life of a structure in a schematic way.

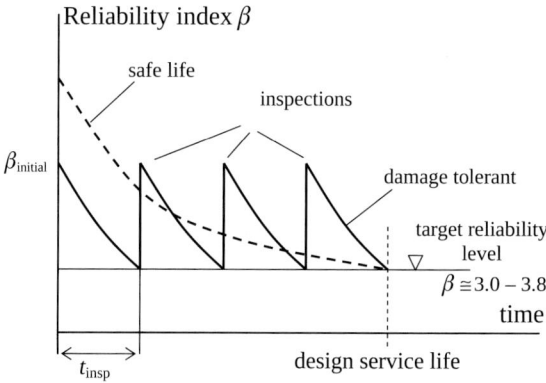

Figure 5.1 – Schematic of fatigue reliability assuming damage tolerant and safe life methods and a failure with high consequence

5.2 STRATEGIES

5.2.1 Safe life

This method is based on the calculation of damage during the structure's design service life using lower bound strength data and an upper bound estimate of the cyclic loading (number of occurrence and intensity). This will provide a conservative estimate of the fatigue life, i.e. the fatigue life will be, with a high level of probability, longer than the design service life.

The safe life method should be used where regular inspections are not possible, e.g. because of a poor accessibility of fatigue critical details or because the owner does not wish to take the commitment and the resulting penalties are acceptable. No regular fatigue inspections need to be specified. However, note that due to other requirements (cleaning, corrosion protection renewal, etc.), maintenance inspections will be made. Major fatigue problems may be detected during maintenance and corrective action taken. Thus, one can say that redundancy is always a desirable feature for structures subjected to cyclic loading.

The economical consequences of this strategy is that the initial cost of the structure can be a higher one, compared to a damage tolerant strategy, but no additional costs are theoretically involved during the design service life (assuming no change of use, proper maintenance, etc.).

5.2.2 Damage tolerant

Because of the scatter in fatigue performance and the possibilities that the structure will stay in service beyond the required minimum life (design service life), the probability that the structure will develop fatigue cracks during its total service life is increased. In such cases, the damage tolerant method can ensure that when fatigue cracking occurs in service, the remaining structure can sustain the maximum working load without collapse until the damage is detected. The damage tolerance method is achieved by one or more of the following:

- selecting details, materials and stress levels that in the case of the formation of cracks would result in a low rate of crack propagation and a long critical crack length;
- providing multiple load paths;

5. RELIABILITY AND VERIFICATION

- providing crack-arresting details;
- providing readily inspectable details for easy regular inspections.

Damage tolerance depends on the level of monitoring the owner is ready/willing/able to apply to the structure and is not automatically ensured by replaceable members. Inspections must be planned to ensure adequate detection and monitoring of damage and to allow for repair or replacement of members. The following factors must be considered:

- location and mode of failure;
- remaining structural strength;
- detectability and associated inspection technique (this should be the largest flaw not likely to be detected rather than the smallest it is possible to find);
- inspection frequency;
- expected propagation rate allowing for stress redistribution;
- critical crack length before repair or replacement is required.

The economical consequences of this strategy is that the initial cost can be a lower one compared to a safe life strategy, but additional costs due to regular inspection have to be considered.

5.3 PARTIAL FACTORS

5.3.1 Introduction

The fatigue verification using the concept of partial factors, as recommended in EN 1993-1-9, is represented by the general relationship (5.1) below.

$$\gamma_{Ff} \cdot \Delta\sigma_E \leq \frac{\Delta\sigma_R}{\gamma_{Mf}} \tag{5.1}$$

where
$\Delta\sigma_E$ stress range, or equivalent stress range, from the action effects corresponding to the total number of applied cycles, N_{tot},
$\Delta\sigma_R$ fatigue strength of the considered construction detail at N_{tot},

γ_{Ff} partial factor on action affects,
γ_{Mf} partial factor on fatigue strength, strategy and consequence of failure.

In this verification, the partial factors γ_{Ff} and γ_{Mf} are taken to cover the dispersions on the side of the actions effects and the determination of the fatigue strength. When these concern structures subject to fatigue loading in particular, the following uncertainties have to be considered:

- the definition of the operating load, and/or the estimation of the stress ranges during the service life, resulting from it;
- the determination of the cycle peaks;
- the presence of flaws in the material and in the connections, i.e. the quality of the used materials and the welded joints;
- the evaluation of the notch effect and thus the process of crack growth in a design detail;
- the applicability of the *Miner's rule* or of linear damage accumulation method (i.e. to get an equivalent constant amplitude stress range).

The partial factors are directly related to the calculation assumptions and the risk assessment of a failure. The vulnerability of people and environment must be reduced to an acceptable residual risk.

The failure due to fatigue is a long-continuous process in which a crack forms in a member, and grows until the remaining cross section can no longer resist the applied static load. The assessment of acceptable residual risk consists of determining whether such a crack can be detected at an early stage, whether the member or the overall structure permits a certain crack, and whether any effective measures to stop the crack growth can be taken.

5.3.2 Action effects partial factor

The EN 1991 assumptions on the action effects result in a recommended value for the partial factor on the action effect side, γ_{Ff}, of 1.0. This value is further repeated in the different EN 1993, EN 1994 and EN 1999 associated Eurocodes as the recommended value. This factor is linked to the lifetime and loading assumptions of the structure or type of structure considered. Fatigue loading assumptions are further detailed in

5. RELIABILITY AND VERIFICATION

sub-chapter 3.1. Regarding lifetimes, below are some indicative design values for different types of structures.

Table 5.1 – Indicative design working life according to EN 1990, including additional information

Design working life category	Indicative design working life (years)	Type of structure
1	10	Temporary structures [1]
2	10 to 25	Replaceable structural parts (e.g. gantry girders, bearings)
3	15 to 30	Agricultural and similar structures
4	50	Building structures and other common structures (e.g. canal lock doors, wind mills, etc.)
5	100	Monumental building structures
5	100	Road and Railway Bridges (EN 1993-2 and EN 1994-2)
-	25 to 50	Runway beams, crane supporting structures (EN 1993-6)
-	30	Towers, Masts and Chimneys (EN 1993-3-1, EN 1993-3-2)

[1] Structures or parts of structures that can be dismantled and re-used should not be considered as temporary.

5.3.3 Strength partial factor

Regarding the partial factor for fatigue strength, compared to the ENV version of the code, a new philosophy was introduced in EN 1993-1-9 to take into account:

- the chosen fatigue verification method (i.e. the strategy chosen) and
- the consequence of failure.

In fact, the fatigue strength factor does not assume a fixed single value, but can be adapted to the characteristics of the structure (e.g. redundancy, regular inspections) as well as the reliability in service and the damage consequences in case of failure. If the structure or details, for

5.3 PARTIAL FACTORS

instance, exhibit fatigue cracking that can be detected and monitored, with predictable crack propagation and limited damage consequences, the data in the EN 1993-1-9 tables 8.1 to 8.10 can be used with the partial factor γ_{Mf} set to 1.0. If these conditions are not fulfilled, for example because the detail cannot be inspected, the partial factor γ_{Mf} value must be increased. EN 1993-1-9 suggests appropriate values for γ_{Mf}, see Table 5.2. Unfortunately, there is currently no link between the proposed table and the guidelines for the choice of consequence class for the purpose of reliability differentiation in EN 1990, annex B, which were presented in section 1.3.4, Table 1.2. The decision criteria in EN 1993-1-9 are not clearly expressed; the values given in Table 5.2 should be regarded therefore as a recommendation only. Each CEN member state has the right to specify appropriate values and criteria in its respective National Annex. The authors propose to add the consequence classes within the table to improve the decision criteria; furthermore, additional explanations are given below.

Table 5.2 – Recommended values for the partial factor γ_{Mf}

Verification method	Consequence of failure	
	CC1 and CC2*	CC3*
	Low consequence	High consequence
Damage tolerance	1.00	1.15
Safe life	1.15	1.35

* see EN 1990, annex B, Table B1

Regarding the decision criteria for choosing the partial factor γ_{Mf}, one has to decide whether the structure or substructure concerned with fatigue failure is damage tolerant or not (the verification method being thus the *safe life method* for the latter). In order to classify as *damage tolerant* the following minimum conditions must be fulfilled:
- during fatigue cracking the possibility for the load transfer exists,
- the critical design details are always accessible and can be inspected

5. Reliability and Verification

(the cases and location of cracking being given in the Tables 8.1 to 8.10 of EN 1993-1-9) and
- crack growth can be stopped (repaired) or the structural member can be replaced.

It can be assumed that the conditions mentioned above are implicitly fulfilled and thus that *damage tolerance* exists when all the following requirements are met together:

- selection of the steel grade according to EN 1993-1-10 to avoid brittle failure, see chapter 6,
- fatigue cracking not occurring from a weld root (e.g. a not inspectable location), but rather from the surface or a weld toe and
- regular inspection and control of the structure by suitably trained and experienced people. The number of these necessary inspections is at least equal to $n_{Insp.} = 3$ over a 100 year service life, i.e. with a constant time interval of 25 years. The inspection interval can also be shorter or variable during the lifetime.

However, there is not yet an holistic approach in the Eurocodes, that is the influence of some parameters on the reliability level are not properly considered or considered at different level along structural design. For fatigue, this is the case for example for fabrication quality and NDT controls, whose importance on reliability level is not known but minimal requirements are fixed. Another example is the case of fatigue cracking in a welded detail under compressive stresses. In this case, fatigue crack propagation rate is likely to be constant (not exponential as in the case of a detail under tensile stresses) and thus the opportunity of detecting cracks during inspections higher before it is considered as having failed. This leads to the criterion *consequence of failure*, for which the following question often arises: when can damage consequences be called "low" and when can they be called "high"? In the case of fatigue cracking in a welded detail under compressive stresses, can it be considered as a failure with low consequence? The authors believe it can, but the general question of estimating the consequence of failure is a point that is always highly debated. The consequences of a failure depend on several parameters, which is partially solved with the definition of the consequence classes from Table 1.2. It remains that things like social consequences are difficult to apprehend, one could think about the cultural heritage of a structure, public confidence in owner or government, etc.

In Figure 5.2, the four options for the partial strength factor value γ_{Mf} are illustrated according to EN 1993-1-9 theoretical reliability background (Sedlacek, 2003), see also Zhao *et al* (1994). It shows schematically the different evolutions of the reliability index β in function of the design strategy option, *damage tolerant* (with inspections) versus *safe life* (without inspections), the failure consequence and corresponding minimum reliability index targets. It can be seen for example the beneficial influence of inspections during the service life of a structure on the reliability index. The reality is however more complex since the reliability index evolution is a function of information on effective loadings, type of inspection, inspection results (no detection, crack detected, crack depth measured), etc.

For a service life of 100 years, as it is typically the case for bridges, a minimum reliability index value of $\beta_{target} = 3.65$ is taken for the category "high consequences" and $\beta_{target} = 0.95$ for the category of "low consequences". For the *damage tolerant* option, it is here assumed that at least three general inspections and/or specific inspections for cracks take place during the planned working life, which corresponds to an inspection time interval of 25 years.

The differences found between the values presented in Figure 5.2 and the recommended values in EN 1990 are due to the fact that Annexes B and C consider a design life of 50 years, usual in the case of buildings. Instead, for bridges, a design life of 100 years is here considered. The corresponding reliability indexes, based on the same annual failure probabilities, have thus been recomputed. The guidance in other EN 1993 and EN 1994 associated Eurocodes may differ from the generic part 1-9 and contain different partial strength factor values (note also that other values may be given in the National Annexes):

- Silos and tanks (EN 1993-4-1, EN 1993-4-2): the recommended value for γ_{Mf} is 1.1.
- Steel and concrete buildings (EN 1994-1-1) and bridges (EN 1994-2): for headed studs, the recommended value for $\gamma_{Mf,s}$ is 1.0.

In the case of orthotropic steel decks details, the recommended value for γ_{Mf} is 1.0 or 1.15 in function of the detail, see Annex B.13 for more information.

5. RELIABILITY AND VERIFICATION

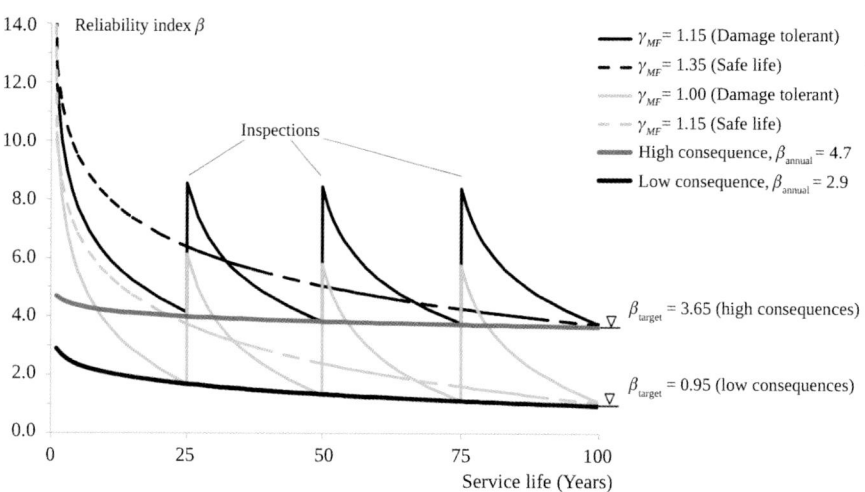

Figure 5.2 – Reliability index, β, as a function of the choice of the verification method (strategy) and the consequence of failure

5.4 VERIFICATION

5.4.1 Introduction

EN 1993-1-9 uses the nominal stress concept to assess the fatigue safety. This means that the design nominal stress range, $\Delta\sigma_{Ed}$ or $\Delta\tau_{Ed}$, resulting from the fatigue action effects, is compared to the design value of the fatigue strength, $\Delta\sigma_{Rd}$ or $\Delta\tau_{Rd}$. Three formats can be distinguished:

1. Verification using the fatigue limit
2. Verification using damage equivalent factors (by convention comparison is made at 2 million cycles)
3. Verification using damage accumulation.

The three cases are described in detail in the following sections. With proper computation of action effects and correct choice of fatigue strength, all three different verification formats can also be applied to a design based on a modified nominal stress range or a geometric stress approach.

The case of fatigue verification under multiaxial stress ranges is then presented in section 5.4.7.

5.4 VERIFICATION

5.4.2 Verification using the fatigue limit

Fatigue tests under variable amplitude stress ranges show, for structural steels, that the life of a structural detail tends to be infinite if all design values of the stress amplitudes $\Delta\sigma_{Ed,i}$ remain below the calculated value of fatigue strength $\Delta\sigma_D/\gamma_{Mf}$ (see section 1.1.3). This observation can be used for the calculation, the general condition becomes:

$$\max\left(\Delta\sigma_{Ed,i}\right) \leq \frac{\Delta\sigma_D}{\gamma_{Mf}} \qquad (5.2)$$

where

    $\max(\Delta\sigma_{Ed,i})$ maximum value of the stress range from the design stress range spectrum $\Delta\sigma_{Ed,i} = \gamma_{Ff}\Delta\sigma_i$,

    $\Delta\sigma_D$ fatigue strength taken as the constant amplitude fatigue limit of the considered construction details (in: $\Delta\sigma_D = 0.74\cdot\Delta\sigma_C$ for $m = 3$),

    γ_{Mf} partial factor for fatigue strength.

The condition specified in relationship (5.2) is not explicitly given in EN 1993-1-9. It arises however as a logical consequence from the acceptance of a fatigue limit at $5 \cdot 10^6$ cycles (total number of cycles from the stress histogram). The application of this verification is valid for all kinds of details subjected to fatigue actions. This verification is on the safe side (conservative) and can be used for instance in the following cases:

- If only the fatigue limit is known or estimated using design assisted by testing (and not the complete fatigue strength curve), see sub-chapter 4.3,
- If the service life is not yet known,
- If the number of cycles during the service life is very high, typically over 100 billions,
- If the shape of the histogram of stress ranges is not known,
- In the context of a preliminary design.

The above fatigue verification is not valid for shear stress ranges as there is no such thing as the constant amplitude fatigue limits (CAFL).

5. RELIABILITY AND VERIFICATION

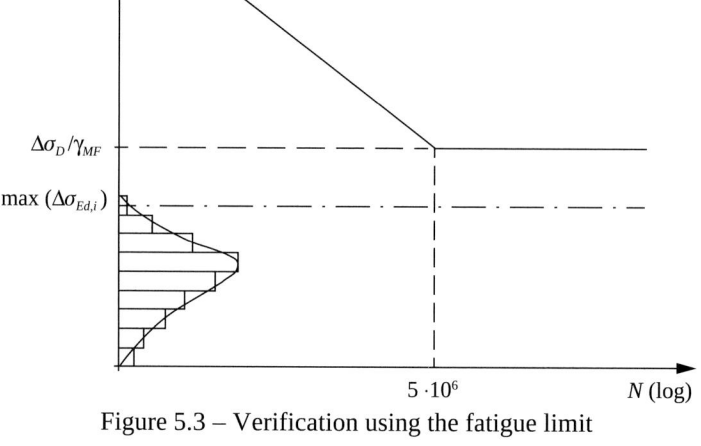

Figure 5.3 – Verification using the fatigue limit

When using verification according to expression (5.2), it should be emphasised that the maximum action effect must be well estimated (max ($\Delta\sigma_{Ed,i}$)) and not underestimated. Also, it should be mentioned that due to a later increase of the service loads, there then may be stress ranges above the CAFL and thus cracking and crack growth may eventually occur. Only one out of ten thousand cycles with a stress range that exceeds CAFL in the histogram has been found to trigger fatigue failure (Fisher et al, 1993). This means that it is not the number of occurrence of stress ranges above CAFL that causes a fatigue crack to start. Rather, it is the fact that this stress range is present in the spectrum and may initiate a crack because there is an uncertainty in the CAFL value (i.e. the uncertainty in the CAFL value is larger than it is in the finite life region). Once fatigue cracking is initiated, it will go through the process of lowering the CAFL and thus enduring a larger portion of damaging cycles in the spectrum and finally it will lead to failure.

This point can be well illustrated by looking at one of the tests conducted by Fisher et al (1993). One of the beams tested experienced a total of 104 billions of cycles of variable amplitude loading. A fatigue crack was found at a transverse stiffener at the end of the test. The damage computations carried out using a detail category 90 give $D_{tot} = 4.29$, see Table 5.3. This is way above unity, but it is normal since we are looking at a test result and damage is computed using the characteristic S-N curve for the detail. When looking at the relative damage according to stress range level, one sees that the higher stress ranges, the only one above the CAFL as well as those just below, do account only marginally for the damage sum. It is the

5.4 VERIFICATION

stress ranges in the middle that account for most of the damage (in this case 30 % to 95 %-quantile). Other tests and damage sum computation hypothesis (slope coefficients, or only one slope, or not considering a cut-off) do not change the conclusions. The highest stress ranges are not the ones doing the damage, they usually account for less than 1 % in the different tests carried out. The higher stress ranges, however, are the ones responsible for initiating the fatigue cracks.

Table 5.3 – Damage analysis of a test result from Fisher *et al* (1993) with only 0.01 % of stress ranges above CAFL

$\Delta\sigma$(MPa)	Fractile	Damage	Relative damage	Cumulative Damage
32.3	1.70 %	0.000	0.0 %	0.0 %
35.6	13.70 %	0.111	2.6 %	2.6 %
38.8	30.70 %	0.242	5.7 %	8.2 %
42.0	49.70 %	0.404	9.4 %	17.7 %
45.2	65.69 %	0.493	11.5 %	29.2 %
48.5	78.69 %	0.565	13.2 %	42.3 %
51.7	86.69 %	0.480	11.2 %	53.5 %
54.9	92.69 %	0.710	16.6 %	70.1 %
58.2	95.69 %	0.422	9.8 %	79.9 %
61.4	97.69 %	0.331	7.7 %	87.7 %
64.6	98.69 %	0.193	4.5 %	92.2 %
67.9	99.29 %	0.134	3.1 %	95.3 %
71.1	99.69 %	0.103	2.4 %	97.7 %
74.3	99.89 %	0.059	1.4 %	99.0 %
77.6	99.99 %	0.033	0.8 %	99.8 %
103.4	100.00 %	0.008	0.2 %	100.0 %
		Total **4.286**		

5. Reliability and Verification

Example 5.1: Detail classification and verification, application to chimney (anchor bolt located at + 11490 mm) (worked example 2)

A detailed description of the details and their fatigue strength is given in Example 4.1.

Partial factors for fatigue loads and strength

One can consider that such details are damage tolerant: the socket can and shall be regularly inspected. Also, since the failure of a bolt does not result in the failure of the chimney and the wind direction is changing, the most loaded bolt and weld zone is not always the same. Thus, the consequences of failure are low. That is:

- Factor for fatigue strength: γ_{Mf} = 1.0 (damage tolerant, low consequence of failure)
- Factor for fatigue loading: γ_{Ff} = 1.0

Fatigue verifications

Verifications are based on nominal stress ranges.
The applied stress range must not exceed the constant amplitude fatigue limit of the respective detail categories.

$$\Delta\sigma_{Ed} < \frac{\Delta\sigma_D}{\gamma_{Mf}}$$

$$\Delta\sigma_E \cdot \gamma_{Ff} = 1.0 \cdot 20.8 < \frac{\Delta\sigma_D}{\gamma_{Mf}} = \frac{29.6}{1.0} = 29.6$$

Socket joint: $\Delta\sigma_E \cdot \gamma_{Mf} = 20.8 \, \text{N}/\text{mm}^2 < 29.6 \, \text{N}/\text{mm}^2$ **SATISFIED**

Bolts: $\Delta\sigma_E \cdot \gamma_{Ff} = 22 \, \text{N}/\text{mm}^2 < 37 \, \text{N}/\text{mm}^2$ **SATISFIED**

Example 5.2: Detail classification and verification, application to chimney (anchor bolt located at + 350 mm) (worked example 2)

A detailed description of the different details composing the bottom assembly is given in Example 4.2.

5.4 VERIFICATION

Loading

Force range in the anchor bolts (computed according to the moment found in example 3.1):

$$\Delta F_t = \frac{2}{n} \cdot \frac{\Delta M_4}{r_s} = 49.4 \text{ kN}$$

Direct stress range in the anchor bolts M60:

$$\Delta\sigma_E = \frac{\Delta F_t}{A_S} = \frac{49400}{2362} = 20.9 \text{ N/mm}^2$$

Number of cycles during the design life

Recall that the design value of the number of load cycles (50 years) is (example 3.5):

$$N_v = 8.6 \cdot 10^8 \text{ cycles}$$

Due to the large number of load cycles ($> 10^8$), the only possible verification to satisfy fatigue design is to do it with the fatigue limit, see section 5.4.2. In other words, this means that one will require stress ranges to stay sufficiently low to have infinite life for all the chimney details.

Fatigue strength of the anchor bolt detail

An anchor bolt in tension corresponds to detail number 14, Table 8.1, see Figure 4.10: this detail is category 50, but since it has a diameter 60 mm, a size effect reduction shall be applied. Thus:

$$\Delta\sigma_{C,red} = \left(\frac{\emptyset_0}{\emptyset}\right)^n \Delta\sigma_C = (30/60)^{0.25} \cdot 50 = 42 \text{ N/mm}^2$$

CAFL for this detail is

$$\Delta\sigma_{D,red} = 0.74 \cdot \Delta\sigma_{C,red} = 31.1 \text{ N/mm}^2$$

Partial factors on fatigue loads and strength

One can consider that such details are damage tolerant since the bottom assembly can and shall be regularly inspected. Also, since the failure of an anchor bolt does not result in the failure of the chimney and the wind

5. RELIABILITY AND VERIFICATION

direction changing, the most loaded anchor bolt and weld zone is not always the same. Thus, the consequences of failure are low. That is:

- Factor for fatigue strength: $\gamma_{Mf} = 1.0$ (damage tolerant, low consequence of failure)
- Factor for fatigue loading: $\gamma_{Ff} = 1.0$

Fatigue verification

Verification is based on nominal or modified nominal stress range.
The applied stress range must not exceed the constant amplitude fatigue limit of the respective detail category.

$$\max\left(\Delta\sigma_{Ed,i}\right) = \Delta\sigma_{Ed} < \Delta\sigma_{D,red}/\gamma_{Mf}$$

Anchor bolts: $\gamma_{Ff}\Delta\sigma_E = 20.9 \text{ N/mm}^2 < 31.1 \text{ N/mm}^2$ **SATISFIED**

Example 5.3: Fatigue verification, application to chimney (welded stiffener to bottom plate at 350 mm) (worked example 2)

A detailed description of the different details composing the bottom assembly is given in Example 4.2. In this verification, we concentrate on the detail welded stiffener to ground plate.

Fatigue strength of the details

It is a tee-butt joint with partial penetration welds, which corresponds to detail number 4 (see Figure 4.11 and Figure 5.4). Recall that there are two potential crack locations for this detail.

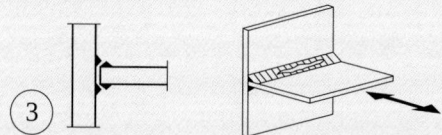

Figure 5.4 – Chimney welded stiffener tee-butt joint (Table 8.5)

Case A: For root cracking:

Under normal stress range, category 36*, and under shear stress range, category 80. Since there are no shear stresses, this verification needs not to be made.

5.4 VERIFICATION

For the normal stress range:

$$\Delta\sigma_C = 36 \text{ N/mm}^2$$

and CAFL: $\Delta\sigma_D = 0.74 \cdot \Delta\sigma_C = 26.6 \text{ N/mm}^2$

However, since it is a detail with a star, one must use the following modified value (see section 4.1.2) for the verification using the constant amplitude fatigue limit:

$$\Delta\sigma_D(\text{at 10 million cycles}) = (2/10)^{1/3} \cdot 1.12 \cdot \Delta\sigma_C^* =$$
$$= (2/10)^{1/3} \cdot 1.12 \cdot 36 = 23.6 \text{ N/mm}^2$$

Case B: For toe cracking:

Direct stress range only and category 80. Thus:

$$\Delta\sigma_C = 80 \text{ N/mm}^2$$

and CAFL: $\Delta\sigma_D = 0.74 \cdot \Delta\sigma_C = 59.2 \text{ N/mm}^2$

Partial factors on fatigue loads and strength

Factor for fatigue strength: $\gamma_{Mf} = 1.0$ (damage tolerant, low consequence of failure)

Factor for fatigue loading: $\gamma_{Ff} = 1.0$

See Example 5.2 for more explanations

Fatigue verification:

Verifications are based on nominal stress ranges.
The applied stress range must not exceed the constant amplitude fatigue limit of the respective detail categories.

$$\max(\Delta\sigma_{Ed,i}) = \Delta\sigma_{Ed} < \Delta\sigma_D / \gamma_{Mf}$$

Case A, root cracking: 16.0 N/mm² < 23.6 N/mm² **SATISFIED**

Case B, toe cracking: 18.7 N/mm² < 59.2 N/mm² **SATISFIED**

5. Reliability and Verification

Example 5.4: Application to chimney, fatigue verification of manhole details (unreinforced or reinforced) (worked example 2)

Fatigue strength of the detail

Unreinforced case:

The manhole is a cut-out made using machine gas cutting with subsequent dressing in order to remove any edge discontinuity. It corresponds to detail number 4, Table 8.1; this detail is category 140. Thus:

$$\Delta\sigma_C = 140 \text{ N}/\text{mm}^2$$

and CAFL: $\Delta\sigma_D = 0.74 \cdot \Delta\sigma_C = 103.6 \text{ N}/\text{mm}^2$

Reinforced case:

In this case, there is also a longitudinal weld (position 2) as well as the end of the stiffeners.

Position 2) this is a manual fillet weld and thus corresponds to detail 5, Table 8.2; this detail is category 100. Thus:

$$\Delta\sigma_C = 100 \text{ N}/\text{mm}^2$$

and CAFL: $\Delta\sigma_D = 0.74 \cdot \Delta\sigma_C = 74 \text{ N}/\text{mm}^2$

Position 3) this is a longitudinal attachment that terminates with an angle $\alpha = 40Y$, which corresponds to detail 2, Table 8.4; this detail is category 71. Thus:

$$\Delta\sigma_C = 71 \text{ N}/\text{mm}^2$$

and CAFL: $\Delta\sigma_D = 0.74 \cdot \Delta\sigma_C = 52.5 \text{ N}/\text{mm}^2$

Partial factors on fatigue loads and strength

Factor for fatigue strength: $\gamma_{Mf} = 1.0$ (damage tolerant, low consequence of failure)

Factor for fatigue loading: $\gamma_{Ff} = 1.0$

See Example 5.2 for more explanations

5.4 VERIFICATION

Fatigue verifications:

The verification is based on a nominal or a modified nominal stress range. In both cases the design value of the applied stress range must not exceed the constant amplitude fatigue limit of the respective detail category.

$$\Delta\sigma_{Ed,mod} < \Delta\sigma_D / \gamma_{Mf}$$

Unreinforced manhole:

$$\gamma_{Ff} \cdot \Delta\sigma_{E,mh1,mod} = 63.2 \text{ N}/\text{mm}^2 < 103.6 \text{ N}/\text{mm}^2 \quad \textbf{SATISFIED}$$

Reinforced manhole:

Position 1) at edge:

$$\gamma_{Ff} \cdot \Delta\sigma_{E,mh2,mod} = 68.9 \text{ N}/\text{mm}^2 < 103.6 \text{ N}/\text{mm}^2 \quad \textbf{SATISFIED}$$

Position 2) weld along reinforcing stiffener:

$$\gamma_{Ff} \cdot \Delta\sigma_E = 43.7 \text{ N}/\text{mm}^2 < 74 \text{ N}/\text{mm}^2 \quad \textbf{SATISFIED}$$

Position 3) end of stiffener:

$$26.5 \text{ N}/\text{mm}^2 < 52.5 \text{ N}/\text{mm}^2 \quad \textbf{SATISFIED}$$

Note: without adding a dynamic damping system to the chimney, in order to increase the logarithm decrement of damping δ value, the computations would show that it is nearly impossible to satisfy fatigue verifications. Even improvements in the strength of the details would not have solved the problem. The reduction of the vibrations due to resonance amplification effects is the key and thus the system must function properly during the whole working life of the structure. It must be periodically inspected (in accordance with EN 1993-3-2 Annex B) and maintained properly to avoid fatigue problems.

5.4.3 Verification using damage equivalent factors

The fatigue verification using damage equivalent factors is the standard procedure. The design value of the equivalent constant amplitude

5. RELIABILITY AND VERIFICATION

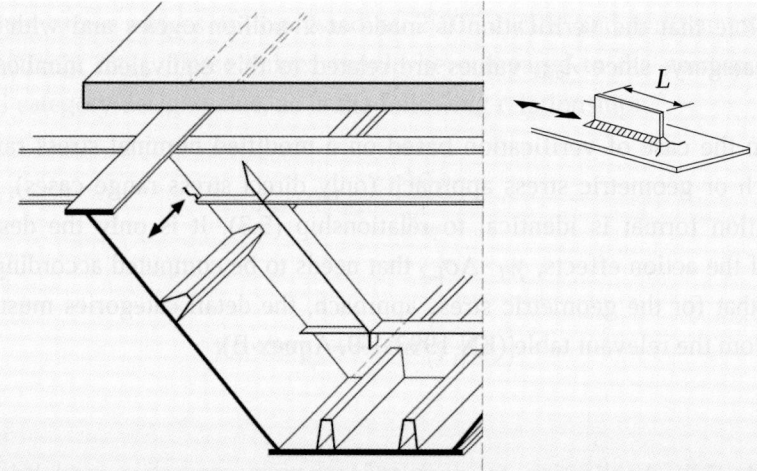

Figure 5.5 – 3D views of a) the road bridge box beam and location of longitudinal attachment, b) its equivalent in the detail category tables

Determination of partial strength factor:

After the failure of the detail, a visible crack at the end of the attachment, the fatigue crack will continue to progress into the web over its full height and then continue into the box flanges. As such, it is a detail of first importance. The consequence of failure of the detail can thus be classified as high, see section 5.3.3. However, the consequences of failure are mitigated by the fact that the detail is located at mid-span and above the neutral axis; as such, it is subjected to compressive stresses. Fatigue cracking will occur at a constant rate (the crack will not grow exponentially) and will take a long time to grow. Furthermore, this type of box is usually inspectable without special equipment (pathway inside the box) and this detail can easily be visually inspected. These conditions fulfill the damage tolerant concept. Finally, using Table 5.2, the value for the partial strength factor can be chosen as $\gamma_{Mf}= 1.15$.

Using the damage equivalent factor already computed in Example 3.2, mid-span P1-P2, the equivalent stress range at 2 millions cycles is computed as follows :

$$\Delta\sigma_{E,2} = \lambda \cdot \Delta\sigma\left(\gamma_{Ff} Q_k\right) = 2.0 \cdot 23.2 = 46.4 \text{ MPa}.$$

The verification is:

$$\Delta\sigma_{E,2} \leq \Delta\sigma_C/\gamma_{Mf} = 56/1.15 = 48.7 \text{ MPa } \textbf{SATISFIED}$$

5.4 Verification

Example 5.6: Application to steel and concrete composite road bridge, verification of the shear connection (worked example 1).

The detail to be checked is a stud welded on the upper flange of the main box-girder bridge. See also Example 3.10 for the calculation of the shear stress ranges $\Delta\tau$ in studs along the bridge and Example 3.9 for the calculation of the direct stress ranges $\Delta\sigma$ on the upper face of the upper flange on which the studs are welded. The stress range $\Delta\sigma$ (respectively $\Delta\tau$) should be multiplied by the partial equivalent damage factor λ (respectively λ_V). The factor λ has been determined in Example 3.2. The factor λ_V is defined in EN 1994-2, clause 6.8.6.2 and determined in the same way as follows:

$$\lambda_v = \prod_{i=1}^{4} \lambda_{v,i} \text{ with}$$

$$\lambda_{v,1} = 1.55 \text{ (road bridge)}$$

$$\lambda_{v,2} = \frac{Q_{ml}}{Q_0}\left(\frac{N_{obs}}{N_0}\right)^{\frac{1}{8}}, N_{obs} = 2 \text{ millions (as in Example 3.2),}$$

$$N_0 = 500000 \; Q_0 = 480\,\text{kN}, \; Q_{ml} = \left(\frac{\sum n_i Q_i^8}{\sum n_i}\right)^{\frac{1}{8}} = 457.4 \text{ kN , so finally}$$

$$\lambda_{v,2} = 1.13 \; \lambda_{v,3} = 1.0$$

$$\lambda_{v,4} = \left[1 + \frac{N_2}{N_1}\left(\frac{\eta_2 Q_{m2}}{\eta_1 Q_{m1}}\right)^8\right]^{\frac{1}{8}} = 1.09$$

Finally it gives the constant factor $\lambda_v = 1.915$ along the bridge.

The corresponding details from EN 1993-1-9 are:

− a crack propagating from the stud weld in the upper flange under direct stress range: Table 8.4, detail 9, $\Delta\sigma_C = 80$ MPa
− a crack propagating in the stud shank under shear stress range: Table 8.5, detail 10, $\Delta\tau_C = 90$ MPa (with $m = 8$).

5. RELIABILITY AND VERIFICATION

The corresponding verifications are respectively $\gamma_{Ff}\Delta\sigma_{E2} = \gamma_{Ff}\lambda\Delta\sigma \leq \Delta\sigma_c/\gamma_{Mf}$ in the upper flange and $\gamma_{Ff}\Delta\tau_{E2} = \gamma_{Ff}\lambda\Delta\tau \leq \Delta\tau_c/\gamma_{Mf,s}$ in the stud shank (see EN 1994-2, clause 6.8.7.2). An additional interaction criterion will be verified later in this Manual, see Example 5.8. It should also be noticed that the first verification under direct stress should only be performed if the steel flange is in tension under the ULS fatigue combination of actions which is the sum of the basic SLS combination of non-cyclic loads and the FLM3, see also Example 3.9.

The load partial factor γ_{Ff} is equal to 1.0 (see recommended value in EN 1994-2, clause 6.8.2).

After the failure of one of the stud shank, the load will be redistributed among the remaining studs and possibly some concrete cracking will occur. This detail is highly redundant and the consequence of failure can thus be classified as low, see section 5.3.3. This type of detail cannot be inspected but still fulfils the damage tolerant concept because of its redundancy as there are many studs and extensive slab cracking should indicate stud failures before complete connection failure occurs. Thus, using Table 5.2, the value for the partial strength factor for stud failure can be chosen as $\gamma_{Mf,s} = 1.0$ (which is the recommended value by EN 1994-2, clause 2.4.1.2(6)).

Note that for the upper main flange, the partial strength factor is still equal to $\gamma_{Mf} = 1.15$, since the same explanations given in Example 5.5 hold true for the upper flange.

Figure 5.6 gives the ratios $\dfrac{\gamma_{Ff}\Delta\sigma_{E2}}{\Delta\sigma_c/\gamma_{Mf}}$ (only in the cross sections with a top flange in tension) and $\dfrac{\gamma_{Ff}\Delta\tau_{E2}}{\Delta\tau_c/\gamma_{Mf,s}}$ along the steel concrete composite bridge. They can be compared to the value 1.0, and since they always stay below unity, it can be concluded that the fatigue verifications related to the shear connectors are satisfied (except for the interaction criterion, not yet checked).

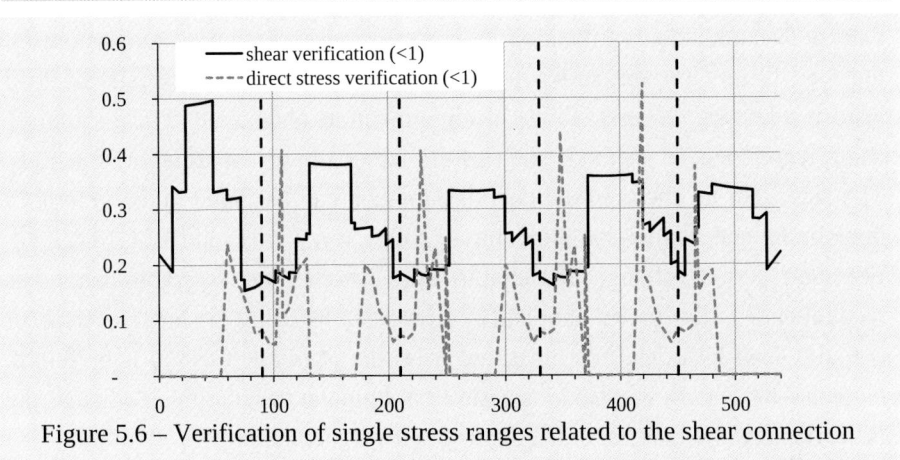
Figure 5.6 – Verification of single stress ranges related to the shear connection

5.4.4 Verification using damage accumulation method

If the service loads are well-known, alternatively, the verification can be performed on basis of the damage accumulation as given in EN 1993-1-9, Annex A. The verification format is then the following:

$$D_d = \sum D_i = \sum \frac{n_{Ei}}{N_{Ri}} \leq D_{max} \qquad (5.7)$$

where

n_{Ei} number of cycles associated with $\Delta\sigma_{Ed,i}$ for band i in the design histogram,

N_{Ri} endurance (in cycles) obtained for a stress range of $\Delta\sigma_{Ed,i}$ considering the partial factors γ_{Ff} and γ_{Mf},

D_{max} limit value of the damage accumulation

In this verification procedure, the stress ranges below the fatigue cut-off limit are generally neglected ($\Delta\sigma_{Ed,i} \leq \Delta\sigma_L/\gamma_{Mf}$). In EN 1993-1-9 the limit value for the damage sum is indicated as $D_{max} = 1.0$. It is assumed here that the safety is assured on the actions side, i.e. in the fatigue load models. Recent experiments have shown, however, that in certain cases, such as post weld treated details, the linear damage accumulation verification according to *Miner's rule* is not always satisfactory (Manteghi and Maddox, 2004). In the new draft version of the IIW recommendations (IIW, 2009) a conservative value of $D_{max} = 0.5$ is therefore recommended. This value is also valid for verifications under both proportional and non-proportional multiaxial stress cases. For practical application, the use of

5. RELIABILITY AND VERIFICATION

damage accumulation verification is, however, an exception as, in most cases, the load history (or the stress range histogram) is not available. If the load history is known, the stress range histogram can be obtained by cycle counting, using recommended methods such as the *Reservoir* or the *Rainflow* methods. The algorithm for the rainflow counting method can be found in any reference book on fatigue, as for example Schijve (2001), IIW (2009) and TGC10 (2006). For short periods of time and simple load histories, the reservoir method is recommended due to its simplicity. However, for long periods of time or complex load histories (e.g. measured data) the *Rainflow* method is preferred. Note that the Rainflow method is easier for computer programming whereas the *Reservoir* method is easier to handle with a hand calculation.

Example 5.7: Application to steel and concrete composite road bridge, damage estimation at end of design service life (worked example 1)

For the longitudinal attachment detail verified in Example 5.5, an estimate of the total damage at the detail at the end of the design service life (100 years) is now made. As an hypothesis, for the whole design life of the bridge, the fatigue load model 4 (with N_{obs} = 2 millions lorries per year) is assumed to be representative for the traffic load crossing the bridge. Note that this traffic (volume and loads) does not correspond to FLM3 used in Example 5.5 but is more representative of a real traffic (see also Example 3.3). Indeed, traffic volume is shared between the 5 different lorries in FLM4. This traffic volume is assumed to have the "long distance" characteristic as given in Table 4.7 from EN 1991-2. Strictly applying this traffic fatigue model, each FLM4 lorry crosses the bridge alone. We simplify the cycle counting by considering only the maximum stress range induced in the studied detail. The results are given in Table 5.4. For some of the lorries, one can see that the stress range is below the cut-off limit for a detail category 56. For the other lorries, the number of cycles to failure under $\Delta\sigma_{Ed,i}$ is computed using the fatigue strength curve, equation (4.1). Each of the damage contributions D_i to the total damage is then computed using equation (5.7). The computed total damage is equal to 1.667, thus above unity, the service life is only 100/1.667 = 60 years. Thus, this example shows that FLM4 is more detrimental than FLM3 and that effectively they are not well calibrated. Specially since the enginner is not rewarded for making the effort of using a larger verification procedure with FLM4 instead of FLM3.

Table 5.4 – Damage sum for FLM4 traffic

FLM4 lorry	Traffic share (%)	n_i/year	$\Delta\sigma_i$ (MPa)	n_i	N_i	$D_i = n_i / N_i$
1	20	400 000	9.7	40 000 000	∞	0.000
2	5	100 000	12.5	10 000 000	∞	0.000
3	50	1 000 000	21.2	100 000 000	69 423 784.58	1.440
4	15	300 000	18.6	30 000 000	∞	0.000
5	10	200 000	20.2	20 000 000	8 839 561 9.25	0.226
Total	100	2 000 000		200 000 000		1.667
				working life =	60.0	years

5.4.5 Verification of tension components

According to EN 1993-1-11, fatigue verification of tension components consist of two separate checks:

1) limitation of maximum stress at serviceability limit states (SLS), with the stress limits values given in Table 5.5. The objective of the maximum stress limit is to keep the corrosion control measures intact, i.e. no cracking of sheaths, hard fillers, no opening of joints etc., and also to cater for uncertainty in the fatigue design.
2) fatigue verification by comparison of stress ranges, as for other details, see EN 1993-1-9, with a special case for infinite fatigue life verification, as explained below.

Table 5.5 – Maximum stress limits f_{SLS} for service conditions (from EN 1993-1-11)

Loading conditions	f_{SLS}
Fatigue design including bending stresses *)	0.50 σ_{uk}
Fatigue design without bending stresses	0.45 σ_{uk}

*) Bending stresses may be reduced by detailing measures

EN 1993-1-11 does not give any indication about either a CAFL or a cut-off limit, see figure below. From literature, we have assumed that there is endurance limit (Cluni et al, 2007). A CAFL might exist but since the fatigue strength of wires, cables is highly depending on: 1) anchorages and thus a combination of tension, bending fatigue as well as fretting fatigue,

5. Reliability and Verification

2) the mean stress value, it cannot be taken into account in the design. Note that in the EN 1993-1-11 (clause 2.4.2) the γ_{Mf} value should be chosen in function of the measures employed to suppress bending effects.

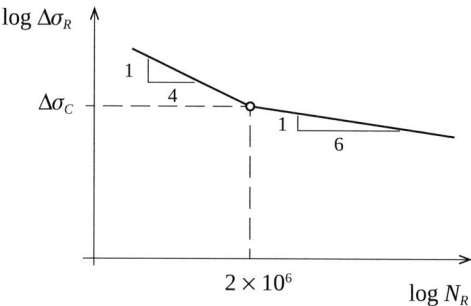

Figure 5.7 – Fatigue strength curves for tension components

Taking an infinite design life approach and adapting the formula for the specific case of cables leads to the following verification relationship:

$$\lambda_{max} \cdot 1.92 \cdot \Delta\sigma\left(\gamma_{Ff} Q_k\right) \leq \frac{\Delta\sigma_C}{\gamma_{Mf}} \qquad (5.8)$$

Where the translating factor is now computed from the S-N curve with a slope $m = 6$ as $\Delta\sigma_C / \Delta\sigma_L \left(10^8 / 2 \cdot 10^6\right)^{1/6} = 1.92$. Practically, this means that all cycles must remain below the value at 100 million cycles, which is a conservative approach. As in equation (5.6), the factor λ_{max} is used in this specific case, even if the S-N curves under constant and variable amplitude loadings are both the same and with two slopes. Finally, note that specific λ factor values should be computed for the S-N curve for cables, but the λ_i, λ_{max} given for welded joints for each type of structure may be used as one can neglect the small differences in the S-N curves slopes (slopes 4 and 6 instead of 3 and 5, and knee point at a slightly different number of cycles). Computations have shown that the influences of such slope changes affect the λ_{max} values by less than 15 % (Maddah, 2013).

5.4.6 Verification using damage accumulation in case of two or more cranes

According to EN 1993-6, clause 9.4.2, in the case of two or more cranes, the combined effect is taken into account by adding the individual damage sums resulting from the cranes acting independently together with the damage index resulting from the cranes acting simultaneously, as follows:

$$\sum_i D_i + D_{dup} \leq 1.0 \qquad (5.9)$$

where

D_i is the damage due to a single crane i acting independently and

D_{dup} is the additional damage due to combinations of two or more cranes occasionally acting together.

Each individual damage D_i, as well as D_{dup}, is to be computed by combining the effects of direct (normal) and shear stress ranges as explained in section 3.7.5.

Example 5.8: Application to runway beam of crane

In this example, the fatigue verification of the runway beam of the crane is performed. The general description, geometry and dimensions are given in section 1.4.4, computation of stress ranges in Example 3.4 and detail classification in Example 4.3.

Structures like the one described are submitted to service conditions requiring their fatigue verification. This verification should be carried out both at the cantilevered supports and their connections with the runway beam. A special care should be taken already at the preliminary design stage regarding potential fatigue effects.

Partial factor for fatigue strength: γ_{Mf} =1.15 (safe life, low consequence of failure). Table 5.6 summarizes the fatigue verification for the details.

Table 5.6 – Summary of the detail fatigue verifications.

Detail	$\Delta\sigma_{E,2}$	$\Delta\sigma_C$ [N/mm²]	$\Delta\sigma_C/\gamma_{Mf}$ [N/mm²]	Verification $\dfrac{\Delta\sigma_{E,2}}{\Delta\sigma_c/\gamma_{Mf}}$	< 1
1	41.2	71	61.7	0.67	OK
2	35.3	125	108.7	0.32	OK
3	28.8	40	34.8	0.83	OK
4	43.5	160	139.1	0.31	OK

Two cranes sharing same runway beams.

5. RELIABILITY AND VERIFICATION

In the case two cranes were acting together, the effects of the crane working isolated and the cranes working together are added according to equation (5.9). The verification in terms of damage accumulation is shown in Table 5.7.

Table 5.7 – Summary of the detail fatigue verifications for two cranes working together

detail	$\Delta\sigma_C$ [N/mm²]	$\Delta\sigma_C/\gamma_{Mf}$ [N/mm²]	Damage due to a single crane i acting independently $\Delta\sigma_{E.2}$ [N/mm²]	D_i	Additional damage due to two cranes acting simultaneously $\Delta\sigma_{E,2dup}$ [N/mm²]	D_{dup}	Total damage of two cranes acting together $2D_i+D_{dup}$	Verification <1?
1	71	61.7	41	0.30	52	0.60	1.19	Not OK
2	125	108.7	35	0.03	44	0.07	0.13	OK
3	40	34.8	29	0.57	36	1.11	2.24	Not OK
4	160	139.1	44	0.03	55	0.06	0.12	OK

Note: for the verification of detail 5. See Example 5.10.

5.4.7 Verification under multiaxial stress ranges

5.4.7.1 Original interaction criteria

Different authors have proposed stress function failure criteria for multiaxial fatigue (Gough *et al*, 1951; Ros and Eichinger, 1950; McDiarmid, 1994; Susmel *et al*, 2001 and Schijve, 2001). For uniaxial normal stress range and a shear stress range, the stress function criterion originally proposed by Gough *et al* (1935) remains the most used one. This criterion is expressed as an elliptical quadrant as follows:

$$\left(\frac{\Delta\sigma}{\Delta\sigma_R}\right)^2 + \left(\frac{\Delta\tau}{\Delta\tau_R}\right)^2 \leq 1.0 \tag{5.10}$$

where
   $\Delta\sigma$ Normal applied stress range,
   $\Delta\tau$ Shear applied stress range,
   $\Delta\sigma_R$ Fatigue strength under normal stress range, expressed at 2 million cycles or at the CAFL level,

$\Delta\tau_R$ Fatigue strength under shear stress range, expressed at 2 million cycles or at the CAFL level.

For biaxial normal stress ranges and a shear stress range, the stress function criterion originally proposed by (Ros and Eichinger, 1950) and based on maximum distortion energy is still in use today for ductile materials (DIN 15018):

$$\left(\frac{\Delta\sigma_x}{\Delta\sigma_{x,R}}\right)^2 + \left(\frac{\Delta\sigma_z}{\Delta\sigma_{z,R}}\right)^2 \pm \frac{\Delta\sigma_x \Delta\sigma_z}{\Delta\sigma_{x,R}\Delta\sigma_{z,R}} + \left(\frac{\Delta\tau}{\Delta\tau_R}\right)^2 \leq 1.0 \qquad (5.11)$$

where
- $\Delta\sigma_x$ Normal applied stress range in direction x,
- $\Delta\sigma_z$ Normal applied stress range in direction z,
- $\Delta\tau$ Shear applied stress range,
- $\Delta\sigma_{x,R}$ Fatigue strength in direction x under normal stress range, expressed at 2 million cycles or at the CAFL level,
- $\Delta\sigma_{z,R}$ Fatigue strength in direction z under normal stress range, expressed at 2 million cycles or at the CAFL level,
- $\Delta\tau_R$ Fatigue strength under shear stress range, expressed at 2 million cycles or at the CAFL level.

These original stress interaction, or stress function, criteria are strictly valid only for proportional cyclic loadings. Also, it should be realised that stress function criteria work only well for crack initation. As soon as a crack has initiated, the stress conditions become different and the crack can take different directions. The prediction of crack growth becomes a more complex problem (Schijve, 2001). Anyway, stress function criteria capture correctly the fatigue behaviour, are simple to use and can be used for verification with proper safety factors. Current standards and recommendations for structural design, such as (IIW, 2009) and the EN 1993 and EN 1994, recommend that S-N curves for normal stress and shear stress be combined with the aid of such stress function interaction criteria. A review of the interaction criteria was carried out by (Backström *et al*, 2004) and more recently by (Sonsino, 2009). One important assumption in the standards is that the criteria under constant amplitude loading directly apply to variable amplitude loading through the use of equivalent stress ranges computed beforehand.

5. RELIABILITY AND VERIFICATION

In the case of proportional loading, the directions of the principal stresses are constant, therefore standards recommend using principal stress range for fatigue verifications (IIW rule), as presented in section 3.7.5. However in the case of non-proportional loadings, or unknown, fatigue verification equations are not identical in the different codes. The phase angle between the loadings (i.e. stresses) influence significantly the fatigue strength. Depending upon the type of multiaxial stresses (only normal, normal and shear), the failure criteria changes (Radaj, 2003).

The rules in the Eurocodes are presented in the next sub-sections.

5.4.7.2 General interaction criteria in EN 1993

In EN 1993-1-9, only the case of combining a uniaxial normal and a shear stress ranges is treated as it is the most common case. The rules for considering combined effects in the fatigue verification, that is when the plane in which a crack is supposed to occur is subjected to a combination of normal and shear stresses (acting proportionally or not), are as follows:

- For proportional loadings, one has only to consider the simultaneous action when this effect is not already included in the constructional detail tables. Effectively, it often occurs that the combination effect is included in the detail category because both normal and shear stresses were acting simultaneously in the tests used to categorize the detail. This is the case, for example, with the details 1 to 9 from Table 8.2, where only the normal stress range has to be considered (the shear stress range in web and flange close to the weld is implicitly included).
- If the normal and shear stresses are synchron (simplest non-proportional case, see section 3.7.5 for definition) and occur at the same location within the detail, or nearly synchron, the maximum principal stress range should be used. This is the case, for example, with the detail 7 in Table 8.4. If the direction of the principal stress continuously changes, then using the maximum values is conservative.
- For non-proportional cases, both synchron and asynchron, the influence of the shear stress range can be neglected when it is less than 15 % of the normal stress range, that is when $\Delta\tau \leq 0.15\Delta\sigma$.
- For the generic case of non-proportional loading, with significant shear stress ranges, $\Delta\tau > 0.15\Delta\sigma$, normal and shear stresses damages

are first assessed separately, then combined using the interaction formula given below (which expresses a damage sum D).

$$D = \left(\frac{\gamma_{Ff}\Delta\sigma_{E,2}}{\Delta\sigma_C/\gamma_{Mf}}\right)^3 + \left(\frac{\gamma_{Ff}\Delta\tau_{E,2}}{\Delta\tau_C/\gamma_{Mf}}\right)^5 \leq 1.0 \quad (5.12)$$

where

$\Delta\sigma_{E,2}$ equivalent constant amplitude normal stress range related to 2 million cycles

$\Delta\tau_{E,2}$ equivalent constant amplitude shear stress range related to 2 million cycles

γ_{Ff}, γ_{Mf} fatigue action effects, respectively fatigue strength safety factors.

The above stress ranges are caused by the fatigue loads specified in the various EN 1991 parts and are given in equations (5.13) and (5.14).

$$\gamma_{Ff}\Delta\sigma_{E,2} = \lambda \cdot \Delta\sigma\left(\gamma_{Ff}Q_k\right) \quad (5.13)$$

$$\gamma_{Ff}\Delta\tau_{E,2} = \lambda \cdot \Delta\tau\left(\gamma_{Ff}Q_k\right) \quad (5.14)$$

where λ is the damage equivalent factor for direct and, by conservative approximation, also for shear stress ranges; $\gamma_{Ff} Q_k$ is the design fatigue load.

The above verification criterion can also be expressed as a straightforward extension of the Miner-rule, that is (allowing for the fact that the slope coefficients of the S-N curves are equal to 3, and 5 under direct, respectively shear stress ranges):

$$D_{d,\sigma} + D_{d,\tau} \leq 1.0 \quad (5.15)$$

where

$D_{d,\sigma}$ is the damage accumulation due to normal stress ranges and

$D_{d,\tau}$ is the damage accumulation due to shear stress ranges.

It should be mentioned that today it is however known that criteria considering individually nominal or structural or hot-spot normal and shear stress ranges and then by adding their damaging increments together do not give a general and sound solution to the multiaxial fatigue behaviour under non-proportional loading, even if they give conservative results for given cases (Sonsino, 2009).

5. RELIABILITY AND VERIFICATION

5.4.7.3 Special case of biaxial normal stresses and shear stress ranges

This case is not explicitly treated in EN 1993-1-9. To explain it, let's take a typical crane runway for top running overhead travelling crane as shown in Figure 5.8a. At the point of wheel load application, the top region of the runway girder (Figure 5.8b) is generally subjected to a stress field comprising local stress components induced by the stress concentrated load, i.e. local transverse pressure $\sigma_{z,local}$ and the local shear stress $\tau_{xz,local}$, in addition to the global stress components σ_x and related shear τ_{xz} due to global bending. For example, the top region near a support of a continuous crane runway is subjected to tensile bending stress σ_x, related shear τ_{xz}, superposed by the local stresses due to wheel load application. And all these stress components may not reach their maxima simultaneously, depending upon the crane runway static system. Furthermore, the stresses $\sigma_{z,local}$ and $\tau_{xz,local}$ are inherently acting out-of-phase, as can be seen on Figure 5.8b.

Thus, under bending and shear stress range, the fatigue crack is likely to propagate vertically, as under local normal stress range, the fatigue crack is likely to propagate horizontally. This case is thus complicated to verify and the interaction between these different loadings is not clearly treated in EN 1993-1-9. However, a criterion such as given in equation (5.11) could be used.

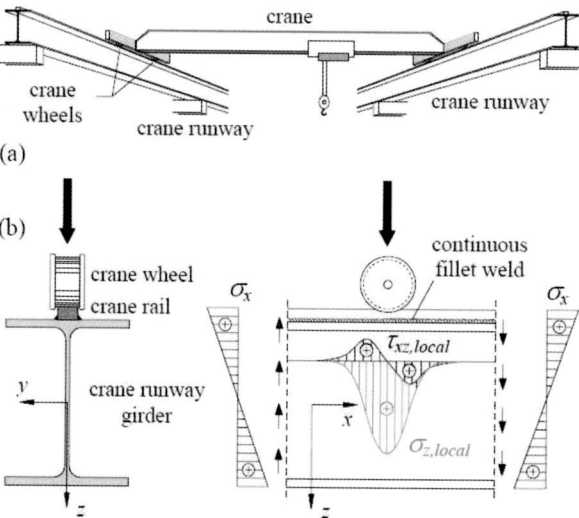

Figure 5.8 – (a) typical crane runway for top running overhead travelling crane, (b) detail of runway under bending and local stresses due to wheel passage

5.4 VERIFICATION

Instead, for crane runways, EN 1993-6, section 9.3.1, requires that the verification is made by taking into account both local and global effects together in equation (5.12). In the above, the effects of global shear stress ranges τ_{xz}, do not need to be accounted for as they are a priori considered negligible with respect to the other contributions. The appropriate detail table and category under bending, shear (Tables 8.1 to 8.5) and local normal stress range (Table 8.10) are to be used. Since there are usually two wheels per crane side, the number of cycles for the local stresses ranges, and thus their damaging effects are to be multiplied by a factor two This results in the following verification:

$$D = \left(\frac{\gamma_{Ff}\Delta\sigma_{x,E,2}}{\Delta\sigma_{x,C}/\gamma_{Mf}}\right)^3 + 2\cdot\left(\frac{\gamma_{Ff}\Delta\sigma_{z,local,E,2}}{\Delta\sigma_{z,C}/\gamma_{Mf}}\right)^3 + \\ +2\cdot\left(\frac{\gamma_{Ff}\Delta\tau_{xz,local,E,2}}{\Delta\tau_{C}/\gamma_{Mf}}\right)^5 \leq 1.0 \quad (5.16)$$

where

- $\Delta\sigma_{x,E,2}$ equivalent constant amplitude normal stress range from bending related to 2 million cycles
- $\Delta\sigma_{x,E,2}$ equivalent constant amplitude local normal stress range related to 2 million cycles
- $\Delta\tau_{xz,local,E,2}$ equivalent constant amplitude local shear stress range related to 2 million cycles
- $\Delta\sigma_{x,C}$ detail category under direct stresses, with fatigue crack propagating vertically
- $\Delta\sigma_{z,C}$ detail category under local normal stresses, with fatigue crack propagating horizontally
- $\Delta\tau_C$ detail category under local shear stresses.

5.4.7.4 Interaction criteria in EN 1994, welded studs

In addition, EN 1994-1-1 gives a special fatigue verification formula for welded shear studs imbedded in concrete (EN 1994-1-1, clause 6.8.7.2). In this specific case, the stresses acting in the detail are:

- a normal stress range in the steel beam flange, to which the stud connectors are welded,

5. Reliability and Verification

– a shear stress range in the weld of the stud connectors due to the composite action effect between the concrete slab and the steel beam.

When the maximum stress in the steel flange to which stud connectors are welded is tensile under the relevant combination, the interaction at any cross section between shear stress range in the weld of stud connectors and the normal stress range should be checked using the interaction formulae given below:

$$\frac{\gamma_{Ff}\Delta\sigma_{E,2}}{\Delta\sigma_C/\gamma_{Mf}} \leq 1.0 \qquad \frac{\gamma_{Ff}\Delta\tau_{E,2}}{\Delta\tau_C/\gamma_{Mf,s}} \leq 1.0 \qquad (5.17)$$

$$\frac{\gamma_{Ff}\Delta\sigma_{E,2}}{\Delta\sigma_C/\gamma_{Mf}} + \frac{\gamma_{Ff}\Delta\tau_{E,2}}{\Delta\tau_C/\gamma_{Mf,s}} \leq 1.3 \qquad (5.18)$$

In this particular case, owing to the fact that the fatigue strength curve for studs in shear is different from the others curves, the computation of the equivalent shear stress range reads:

$$\gamma_{Ff}\Delta\tau_{E,2} = \lambda_v \cdot \Delta\tau\left(\gamma_{Ff}Q_k\right) \qquad (5.19)$$

where λ_v is the damage equivalent factor for studs under shear stress range, see section 3.2.3 and also EN 1994-2 for bridges. For buildings, since no damage equivalent factors are given, EN 1994-1-1 specifies to use the damage accumulation verification format, with proper account for the different fatigue strength curve slopes.

Example 5.9: Application to steel and concrete composite road bridge, verification of the shear connection interaction criteria (worked example 1)

This example concludes Example 5.6 with the third verification for interaction according to relationship (5.18). This relationship should only be used to check details in the bridge cross sections where the upper flange with the welded shear connectors is in tension.

Figure 5.9 gives this additional check as a complement Figure 5.6 from Example 5.6. In Figure 5.9, the results from the check using relationship (5.18) are added. It can be seen that the interaction criterion is also satisfied for the composite bridge example of this Design Manual.

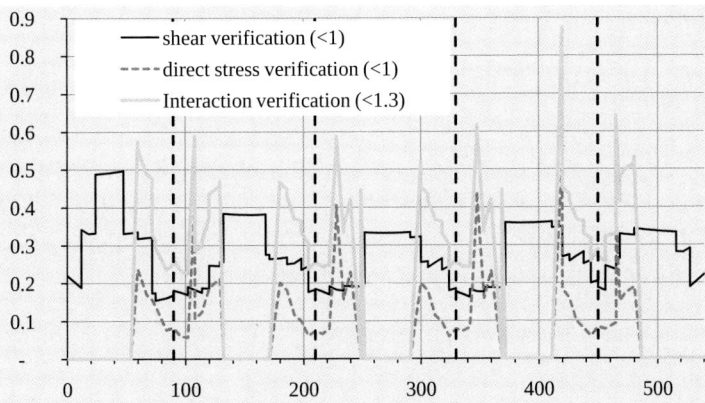

Figure 5.9 – Verification of the interaction criterion for the shear connection

Example 5.10: Application to crane supporting runway beam, verification under multi-axial stress ranges of details 2 and 5 (see Figure 4.12).

The section with the highest normal stresses is at mid-span, and even tough the stress ranges due to bending are compressive, it shall be checked using the full stress ranges since it is a welded beam.

The verifications are carried out according to EN 1993-6, that is using verification condition (5.16) and the appropriate detail categories:

$$D = \left(\frac{\gamma_{Ff}\Delta\sigma_{x,E,2}}{\Delta\sigma_{x,C}/\gamma_{Mf}}\right)^3 + 2\cdot\left(\frac{\gamma_{Ff}\Delta\sigma_{z,local,E,2}}{\Delta\sigma_{z,C}/\gamma_{Mf}}\right)^3 + 2\cdot\left(\frac{\gamma_{Ff}\Delta\tau_{xz,local,E,2}}{\Delta\tau_{C}/\gamma_{Mf}}\right)^5 \leq 1.0$$

– Computations of the stress ranges:

The normal stress range from bending has been determined previously, see Example 3.4. The shear stress range from bending is considered negligible.

The local stresses are determined according to EN 1993-6 chapter 5.7. The effective length, l_{eff} is computed as:

$$l_{eff} = 3.25\cdot\left(\frac{I^*}{t_w}\right)^{1/3} = 3.25\cdot\left(\frac{189826}{8}\right)^{1/3} = 93.4 \text{ mm}$$

5. Reliability and Verification

The resulting normal local stresses, are:

$$\Delta\sigma_{z,local,E,2} = \sigma_{z,local,E,2} - 0 =$$

$$= \frac{Q_{E,2}}{t_w \cdot (l_{eff} + 2r)} = \frac{34 \cdot 10^3}{8 \cdot (93.4 + 2 \cdot 24)} = 30.4 \text{ N/mm}^2$$

The local shear stress ranges can be computed as follows:

$$\Delta\tau_{xz,local,E,2} = \tau_{xz,local,E,2\max} - \tau_{xz,local,E,2\min} = 2\cdot\tau_{xz,local,E,2} =$$

$$= 2 \cdot 0.2 \cdot \sigma_{z,local,E,2} = 0.4 \cdot 30.4 = 12.2 \text{ N/mm}^2$$

- Detail categories for detail 2:

For stress ranges from bending, the KSN is a rolled product which corresponds to detail 1 from Table 8.1, that is cat. 160.

For local normal stress ranges, since it is fillet welds, the appropriate detail category from Table 8.10 is cat. 36.

For the fillet weld shear it corresponds to detail 8 in Table 8.5, that is cat. 80.

- Detail categories for detail 5:

For stress ranges from bending, the HEA is a rolled product which corresponds to detail 1 from Table 8.1, that is cat. 160.

For local normal stress ranges, since it is a rolled H-section, the appropriate detail category from Table 8.10 is cat. 160.

For the web in shear it is conservatively taken as detail 8 in Table 8.5, that is cat. 80.

The verification for detail 2 follows:

$$D = \left(\frac{1.0 \cdot 35.3}{160/1.15}\right)^3 + 2 \cdot \left(\frac{1.0 \cdot 30.4}{36/1.15}\right)^3 + 2 \cdot \left(\frac{1.0 \cdot 12.2}{80/1.15}\right)^5 = 1.82 \leq 1.0$$

This verification is not satisfied.

The verification for detail 5 follows:

$$D = \left(\frac{1.0 \cdot 28.6}{160/1.15}\right)^3 + 2 \cdot \left(\frac{1.0 \cdot 30.4}{160/1.15}\right)^3 + 2 \cdot \left(\frac{1.0 \cdot 12.2}{80/1.15}\right)^5 = 0.03 \leq 1.0$$

It can be seen that this verification is largely satisfied.

Chapter 6

BRITTLE FRACTURE

6.1 INTRODUCTION

Under specific service conditions, steel structures have shown that steel could be sensitive to a fracture type called brittle fracture, especially if they contained welds. Several catastrophic cases, such as the tank failure in Boston in 1919, the T-2 tankers and Liberty ships during the period 1942-1952 or the Hasselt Bridge in Belgium in 1938 have been reported (Barsom and Rolfe 2000) (Akesson, 2008). In all these cases, the material used in these structures had met all existing tensile and ductility requirements and the fracture remain mysterious for some time. It is these failures and the research done to explain them that led to the development of linear elastic fracture mechanics, in particular by Georges R. Irwin (Landes, 2000).

In order to measure steel resistance against brittle fracture, a test called the "Charpy test" has been introduced. This test consists in breaking a specimen with a sharp notch under dynamic loading (impact test) at a prescribed temperature. One measures with this test the energy absorbed by the specimen during its fracture. This test carried out at different temperatures allows for the evaluation of impact resistance of a notched specimen and its sensitivity to brittle fracture. The conditions to carry out a Charpy test are specified in EN 10 045-1 (EN 10 045-1, 1990), which uses a standard ISO specimen geometry with a V-notch, see Figure 6.1. The test is thus often referred to as the Charpy V-notch test, abbreviation CVN, and it is the most commonly used.

6. BRITTLE FRACTURE

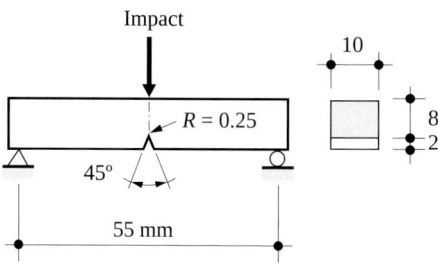

Figure 6.1 – Standard ISO specimen geometry with a V-notch for the Charpy test (TGC 10, 2006)

The CVN test is a cheap and easy to perform mechanical test that gives an indication on the amount of energy a material can absorb during fracture, A_V, expressed in Joules. Figure 6.2 shows examples of CVN impact tests results for a steel S355.

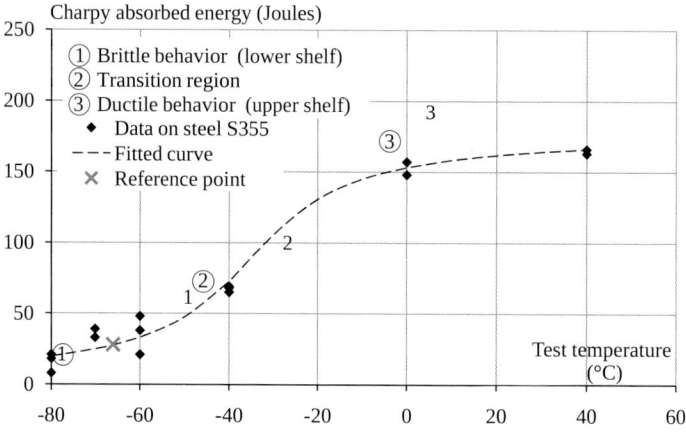

Figure 6.2 – Example of Charpy impact test results at different temperatures and transition curve (S355N steel) (Banz and Nussbaumer, 2001) - 1) Brittle behaviour (lower shelf); 2) Transition region; 3) Ductile behaviour (upper shelf)

The absorbed energy is a measure of the toughness of a given material and acts as a tool to study toughness temperature dependency of materials. Steel shows a strong temperature dependency as can be seen in Figure 6.2, with a transition from a ductile to a brittle behaviour with decreasing test temperature. In the lower shelf region, absorbed energy values are typically lower than 15 J.

If the tests are carried out at low strain rates (static tests), one can observe a shift of the transition curve to the left as shown schematically in

Figure 6.3, i.e. at the same temperature, the material can absorb more energy before fracture. Figure 6.3 shows that toughness of structural steels is a function of both temperature and loading rate.

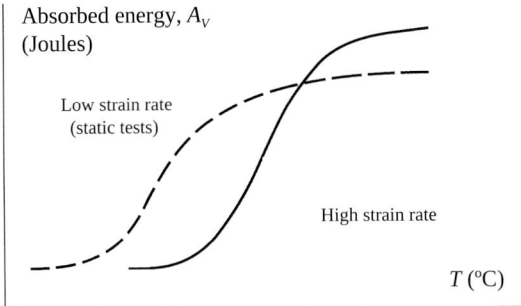

Figure 6.3 – Schematic difference between static and impact tests transition curves for a structural steel (TGC 10, 2006)

6.2 STEEL QUALITY

By measuring the absorbed energy, it is possible to differentiate one steel from another and to define its steel quality (e.g. in addition to the steel grade). The designation of steel qualities are given in the relevant parts of EN 10 025, where the steel quality classes are defined by fixing a minimum or guaranteed value obtained from the Charpy V-notch impact tests at a specified test temperature. The quality classes are codified by adding letters and figures to the steel grade, as given in the Table 6.1 below.

Table 6.1 – Definition of steel quality according to European codes

Code	Notation	Absorbed energy, Joules	Test temperature, °C
EN 10 025-2 Non alloy structural steels	J	27	
	K	40	
	R		+ 20
	0		0
	2		- 20
EN 10 025-3 Fine grain structural steels	-	40	- 20
	L	27	- 50

For example, a steel quality **K2** means a guaranteed value of 40 joules at −20 °C (test temperature). Increasing steel qualities that are usually chosen can be defined according to the following list: JR, J0, J2, K2 for non alloy structural steels, N or NL for fine grain structural steels (N for normalised) and finally M or ML for thermo-mechanical structural steels. Note also that the steel weldability increases from the quality JR to J2. In terms of weldability and toughness, the best steels are the thermo-mechanical steels. Their lower carbon content allows for a reduction or even omission of preheating before welding (SED, 2004).

It should be here emphasised that the absorbed energy at a given test temperature is not a direct indication of the brittle fracture risk in a steel structure because the specimen geometry (with its notch) and loading conditions (impact) do not correspond to the real conditions. Everything considered, the risk of brittle fracture in a structure depends upon the following parameters:

- Service temperature,
- Loading strain rate,
- Flaws (type, size, shape),
- Dimensions of the structural member, also known as thickness or constraint effect,
- Residual stresses,
- Ratio stress level to yield strength, σ/f_y.

All these parameters are taken into account in the fracture concept in EN 1993-1-10, which will be developed in the next sub-chapter. But before, one needs a mean to correlate the different fracture toughness tests results in order to be able to predict the real structural behaviour and fracture load.

6.3 RELATIONSHIP BETWEEN DIFFERENT FRACTURE TOUGHNESS TEST RESULTS

As seen in the previous sub-chapter, the toughness properties vary with temperature. This observation is valid for any type of fracture toughness test. Figure 6.4 gives a complete overview of the function of the toughness-temperature dependency, for which the following regions are distinguished:

- lower shelf region, where the load-deformation characteristic of test pieces in tension show brittle behaviour and linear elastic fracture mechanics may be used featuring stress intensity factors K_{IC} as toughness values,

6.3 RELATIONSHIP BETWEEN DIFFERENT FRACTURE TOUGHNESS TEST RESULTS

- transition region with partial plastic deformations where modified linear elastic fracture mechanics may be used and the temperatures T_{gy} signifies the point where general yield in a net-section (e.g. for a plate with bolt holes) occurs before fracture,
- upper shelf region, where the load-deformation characteristic of test pieces in tension show full ductile behaviour and non linear elastic plastic fracture mechanics applies.

As mentioned in Figure 6.4 under number 1, the material fracture mechanism goes from a brittle to a ductile mode with increasing temperature. At low temperature a brittle, cleavage mode, with fracture through the material grains and little deformations even at the microscopic level is observed. At higher temperatures a ductile tearing, shear mode, with extensive plastic deformations and dipples is observed (Broek, 1986) (Barsom and Rolfe, 2000).

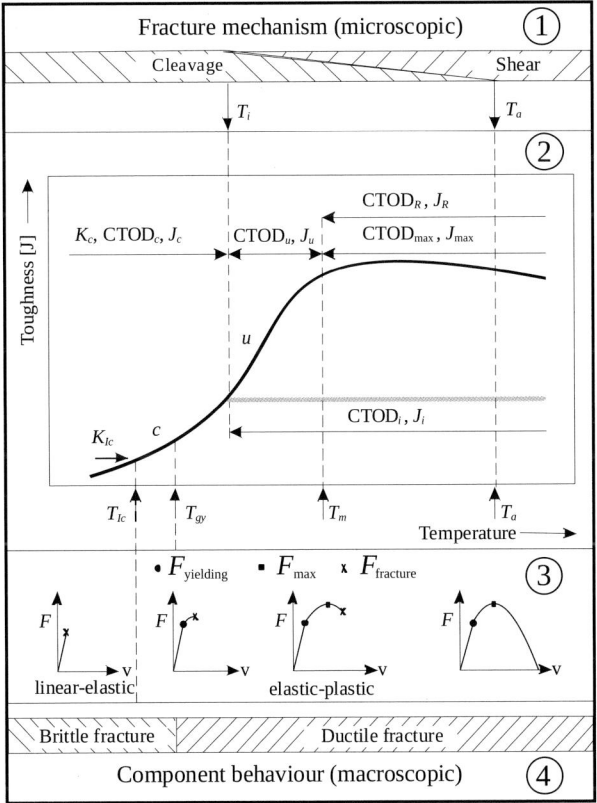

Figure 6.4 – Toughness-temperature curve and related load-deformation curves for tension members using various parameters for toughness properties for ferritic steels (Sedlacek *et al*, 2002)

6. BRITTLE FRACTURE

Quantitative toughness properties of steel in general are determined by standardized fracture mechanics tests, which are significantly more expensive than CVN impact tests. The results of these tests give, as written in Figure 6.4, number 2, different measures and units for the fracture toughness: K in N/mm$^{3/2}$, CTOD in mm, and J in N. For information on CTOD (Crack Tip Opening Displacement) and J (J-integral), see literature as for example (Barsom and Rolfe, 2000). Since correlations between the fracture parameters depend upon the region, an index is added to precise the region in which the fracture toughness was measured: c for the lower shelf, u for the transition region and R or max for the upper shelf.

The stress-strain relationships corresponding to each region are shown in Figure 6.4, number 3. In structural codes, the design rules for achieving sufficient mechanical resistance and stability of structural members and structures are based on continuum mechanics and tests that are carried out in laboratories at room temperature. The assumption behind the design rules is that upper shelf toughness behaviour and ductile stress-strain-behaviour govern the performance of test pieces (Sedlacek *et al*, 2002). Therefore it is necessary to avoid brittle fracture by an appropriate choice of material, i.e. a steel with sufficient toughness.

In this respect, the most important parts of the toughness-temperature curve are the lower shelf and beginning of the transition regions, i.e. where the transition from brittle to ductile fracture occurs (temperature T_{gy}, see Figure 6.4). Fracture-mechanics-type specimen tests results show the same type of transition curve as previously seen with the CVN impact tests (Figure 6.2) but the transition does not occur at the same temperature.

The most interesting correlations are the one between small-scale inexpensive fracture-toughness tests (Charpy V-notch impact specimen) and the larger, more expensive fracture-mechanics-type specimens (K, CTOD, J). Many studies have been made on this topic and numerous empirical correlation expressions over the last decades have been proposed. They correlate values between different impact tests, between impact and static tests, for the transition-temperature region, for the upper shelf region, etc. For this problem, that is correlation between CVN and K-values in the transition-temperature region, expressions have been proposed, for example, by Barsom (Barsom, 1975), Sanz (Sanz, 1980) and

6.3 RELATIONSHIP BETWEEN DIFFERENT FRACTURE TOUGHNESS TEST RESULTS

more recently Wallin (Wallin, 1994). These expressions make the link between dynamic values obtained from CVN tests and static K-values. In the cases of Sanz and Wallin, the correlations are called two-stage CVN-K_{Id}-K_{Ic} correlations. It means that estimates of the static value of the fracture toughness, K_{Ic}, at any strain rate can be predicted by using CVN data in conjunction with an impact correlation relationship to get dynamic fracture toughness, K_{Id}, values and then shifting the curve to lower temperature (to account for strain rate effects). The reference points for the temperature shift on each curve (i.e. on the absorbed energy-temperature curve and on the toughness-temperature curve) are usually taken at an absorbed energy value of 27 J and at a fracture toughness of 3160 N/mm$^{3/2}$ (100 MPa·m$^{1/2}$). These values are conventionally located at the beginning of the transition region, see Figure 6.2 and Figure 6.4 (point with coordinates T_K, K_K). The correlation expressions may include different material properties such as the yield stress, the elasticity modulus, the absorbed energy or the reference point for the temperature, etc. In EN 1993-1-10, the correlation expression used is the one proposed by Wallin (Wallin, 1994), also called master curve approach, following previous work by Sanz (Sanz, 1980); it is given in equation (6.1) below.

$$K_{mat} = 20 + \left[70\exp\left(\frac{T-T_{K100}}{52}\right)+10\right]\left(\frac{25}{b_{eff}}\right)^{1/4} \cdot \left(\ln\left(\frac{1}{1-P_f}\right)\right)^{1/4} \quad (6.1)$$

where
- K_{mat} fracture toughness at temperature T (equivalent to K_{Ic} if plane strain conditions are respected),
- T temperature in °C,
- T_{K100} temperature at which the toughness will not be less than 100 MPa·m$^{1/2}$,
- 25/b_{eff} thickness and constraint effects along the crack front, b_{eff} being related to the crack depth and the plate dimensions in which it grows, see EN 1993-1-10 background document, annex A (JRC, 2008). By simplification, b_{eff} can be conservatively taken as twice the plate thickness,
- P_f failure probability.

6. BRITTLE FRACTURE

In equation (6.1) above, the temperature shift can be directly included with the expression given in equation (6.2).

$$T_{K100} = T_{KV27} - 18 \qquad (6.2)$$

where

T_{KV27} temperature at which a minimum energy A_V will not be less than 27 J in a CVN impact test.

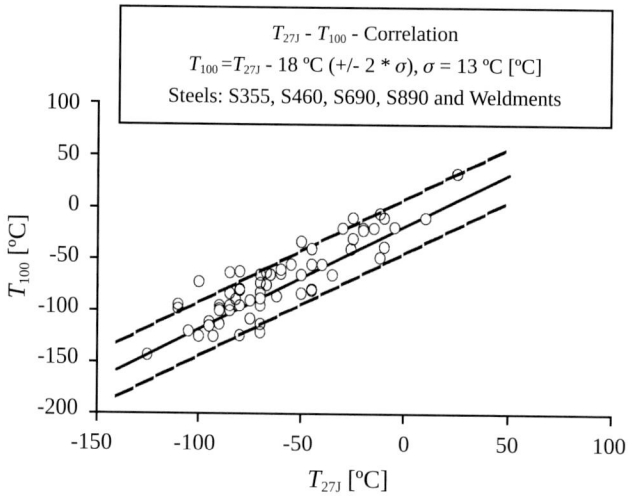

Figure 6.5 – Temperature shift between CVN and toughness tests (JRC, 2008)

The correlation of temperature shift between CVN and toughness tests values is given in Figure 6.5. A constant shift of -18°C is found to be valid for different structural steels and their weldments. A similar approach with a temperature shift can be used to account for other effects such as cold working or inhomogeneity of the material properties in the thickness (core versus surface).

As an example, Figure 6.7 shows the predicted static fracture toughness, K_{mat}, obained using the CVN test results for the S355 steels presented in Figure 6.2. In this case, applying relationship (6.2) gives: $T_{K100} = T_{KV27} - 18 = -67 - 18 = -85$ °C. The mean curve as well as 5 % and 95 % probability curves are drawn. The upper shelf region is not shown since the Wallin correlation is a lower shelf-transition correlation.

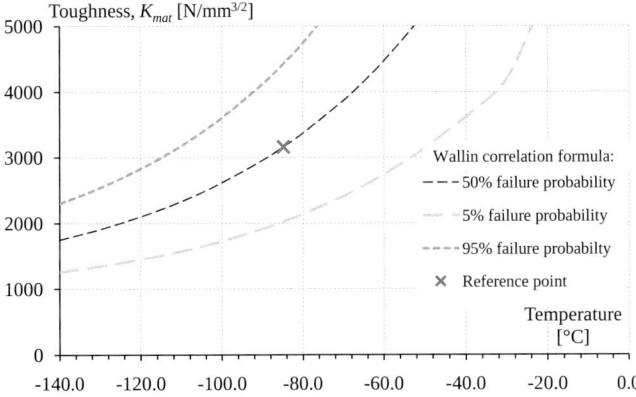

Figure 6.6 – Predicted static fracture toughness for the S355 N steel from (Banz and Nussbaumer, 2001)

6.4 FRACTURE CONCEPT IN EN 1993-1-10

6.4.1 Method for toughness verification

When a steel structure is built, it shall satisfy the requirements for the execution of steel structures (from the codes EN 1090, execution and ISO 5817, welding) but will however contain flaws that are within the required tolerances. Since it is possible that fatigue cracks develop from these flaws, or that undetected flaws may be present, brittle failure shall be excluded by a proper choice of the material.

EN 1993-1-10 provides a method for selecting such a material, for the different possible applications and service conditions (characterized by a minimum reference temperature and stress level). The fracture concept for choosing the proper material is to verify that the material has sufficient toughness to resist the design loads without fracture. In this verification, structural members are assumed to have flaws, which is unavoidable even if within fabrication tolerances. In the general delivery conditions, every structural steel is required to have a minimum Charpy value at a test temperature given by the code (the temperature and value depends upon grade and quality). As seen in the previous section, this requirement can be used to compute the corresponding toughness of the material at any temperature using equation (6.1). Once the toughness is known, one could use fracture mechanics concepts to compute the failure load of the structural member as suggested in EN 1993-1-10. However, the code

6. BRITTLE FRACTURE

developers did not write it with this in mind and thus the code only gives generic principles. Instead, the code developers intention were to simplify the practicing engineer work. Fracture mechanics computations were carried out once for all to provide tables for the choice of material. This work resulted in tables where maximum allowable thicknesses for the different steel grades and qualities are given in function of the different influencing parameters: type of structure and its use, minimum service temperature, and loading conditions (maximum load, loading rate and residual stress level), see sub-chapter 6.5 for more explanations.

In this book, only the general verification method is explained, leaving aside refinements such as the use of the R6 Fracture assessment diagram from BS PD 7910 (BS7910, 1999) or the dependency of the material yield stress on temperature and thickness. Detailed information on the complete verification method can be found in the code background document (JRC, 2008). The verification method uses fracture mechanics, which is not presented here, the reader being referred to the specialised literature, for example (Broek, 1986) (ASM, 1996) (Barsom and Rolfe, 2000). Knowledge in fracture mechanics is however not needed to understand the verification method, which follows the usual verification principle: $E_d \leq R_d$. In this case, it is performed by comparing K-values (stress intensity factors) of, on one side, design values of fracture mechanical action effects K_{Ed} with, on the other side, design values of fracture mechanical resistance K_{Rd}, see equation (6.3).

$$K_{Ed} \leq K_{Rd} \tag{6.3}$$

The design values are chosen from statistical distributions in such a way that the reliability required for ultimate limit state verifications is achieved. The verification is based on the following conservative assumptions:

- the temperature $T_{min,d}$ of the structural member, which is the leading action in this particular ultimate limit state, attains its minimum value,
- the structural member has a crack-like flaw at the point of maximum stress concentration (hot spot) with the size a_d (e.g. design value of crack depth, larger than the flaws that are within the required tolerances, see section 6.4.3),
- the structural member has residual stresses from fabrication and those are considered similarly as an external load, see section 6.4.4,
- the structural member is subjected to permanent and variable loads accompanying the leading action $T_{min,d}$, see section 6.4.4.

6.4 FRACTURE CONCEPT IN EN 1993-1-10

The design situation combining all of the assumptions listed above is accidental. By using a K-value format for the verification, see equation (6.3), it is possible to take advantage of the correlation between CVN and fracture toughness K-values. Thus, since the CVN values specified in the delivery standards for steels are used, the steels may be selected without the need for expensive toughness data determined for each specific project.

The resistance side in the verification, K_{Rd}, corresponds to the Wallin correlation, equation (6.1), which becomes equation (6.4) below when assuming a low failure probability (typically less than 1 %; in this case the last part of the equation (6.1) falls out).

$$K_{Rd} = 20 + \left[70\exp\left(\frac{T_{min,d} - T_{K100}}{52}\right) + 10\right]\left(\frac{25}{b_{eff}}\right)^{1/4} \quad (6.4)$$

where b_{eff} is function of the plate thickness t and crack constraints conditions and is expressed in mm. Thus, there is a direct link between toughness and plate thickness. Note that the safety element in the verification has been calibrated accounting for the above and is a temperature shift, see section 6.4.2. Putting equation (6.4) into the verification relationship (6.3) and rewriting it in function of temperature leads to the following relationship (6.5).

$$-52\ln\left[\frac{(K_{Ed} - 20)\left(\frac{b_{eff}}{25}\right)^{1/4} - 10}{70}\right] + T_{min,d} \geq T_{K100} = T_{KV27} - 18 \quad (6.5)$$

EN 1993-1-10 uses equation (6.5) for safety verifications, however it is presented in a different manner which is now explained.

6.4.2 Method for safety verification

In EN 1993-1-10, the leftmost term of the relationship (6.5) is called ΔT_σ, that is (equation (6.6)):

6. BRITTLE FRACTURE

$$\Delta T_\sigma = -52 \ln \left[\frac{(K_{Ed} - 20)\left(\dfrac{b_{eff}}{25}\right)^{1/4} - 10}{70} \right] \quad (6.6)$$

It is defined as the temperature shift resulting from: stresses, crack like flaws and member shape and dimensions. For safety verification, the relationship (6.5) is given in the form of equation (6.7) below.

$$T_{Ed} \geq T_{Rd} \quad (6.7)$$

where T_{Ed} is a reference temperature. As mentioned before, other material influences can in addition be introduced as temperature shifts. T_{Ed} includes all input values by taking them into account by temperature shifts as expressed by equation (6.8).

$$T_{Ed} = T_{min,d} + \Delta T_r + \Delta T_\sigma + \Delta T_R + \Delta T_{\dot{\varepsilon}} + \Delta T_{\varepsilon pl} \quad (6.8)$$

The input values are (see also Figure 6.7):
- the lowest air temperature $T_{min,d}$ with a specified return period, see EN 1991-1-5
- radiation losses ΔT_r of the structural member (usually $\Delta T_r = -5\ °C$ as given in EN 1991-1-5, but can be defined in the National Annexes),
- the influence of shape and dimensions of the member, imperfection from crack, and stress σ_{Ed}, resulting in ΔT_σ.
- an additive safety element ΔT_R, by which $T_{min,d}$ is shifted, is introduced to achieve sufficient reliability for the verification.
- the influence of strain rate $\Delta T_{\dot{\varepsilon}}$,
- the influence from cold forming $\Delta T_{\varepsilon pl}$, if the cold forming operation is applied after ΔT_{KV27} has been defined (i.e. the CVN tests carried out, or the CVN specified value, are based on the material properties before cold forming).

The resistance side contains solely the test temperature value T_{KV27} and the temperature shift of 18 °C from the toughness correlation. The additional safety element ΔT_R is obtained from a calibration of the procedure to large scale tests database (reliability index β equal to 3.8). The large database used contains tests on various steel grades, various welded attachments including

6.4 Fracture Concept in EN 1993-1-10

local residual stresses and also cracks a_d produced by artificial initial cracks grown by subsequent fatigue loading, see next section and (JRC, 2008).

$$K_{Ed} \leq K_{Rd} \longrightarrow \text{Transformation} \longrightarrow T_{Ed} \geq T_{Rd}$$

$T_{Ed} \geq T_{Rd}$

Action side ← → Resistance side

$$T_{Ed} = T_{min,d} + \Delta T_r + \Delta T_\sigma + \Delta T_R \left[+\Delta T_{\dot{\varepsilon}} + \Delta T_{\varepsilon pl} \right]$$

- Lowest air temperature in combination with q_{Ed}

  $T_{min,d} = --25\ °C$

- Radiation loss

  $\Delta T_r = --5\ °C$

- Influence of stress, crack imperfection and member shape and dimension

$$\Delta T_\sigma = -52 \cdot \ln \left[\frac{(K_{Ed} - 20)\left(\frac{b_{eff}}{25}\right)^{1/4} - 10}{70} \right] \quad [°C]$$

- Additive safety element

  $\Delta T_R = -+7\ °C$ (with $\beta = 3.8$)

∞ Influence of material toughness

$T_{100} = T_{27J} - 18\ [°C]$

May be supplemented by the following:

- Influence of the strain rate

$$\Delta T_{\dot{\varepsilon}} = \frac{1440 - f_y(t)}{550} \left(\ln \frac{\dot{\varepsilon}}{\dot{\varepsilon}_0} \right)^{1.5} \quad [°C]$$

- Influence from cold working

  $\Delta T_{\varepsilon pl} = --3 \cdot DCF\ [°C]$

Influence from cold working: DCF = Degree of Cold Working [%].
See JRC (2008) and for hollow sections JRC (2012).

Figure 6.7 – Verification scheme based on temperatures, with example values for temperature shifts (Sedlacek *et al*, 2002)

6. BRITTLE FRACTURE

6.4.3 Flaw size design value

For a given detail, the accidental existence of a flaw (modelled as an initial crack with size a_0), that should normally have been detected and repaired during welding inspection, is assumed. The initial crack size is assumed to depend on the plate thickness, the initial crack depth is given in equations (6.9) and (6.10) and shown in Figure 6.8.

$$a_0 = 0.5 \cdot \ln\left(1 + \frac{t}{t_0}\right) \text{ for } t < 15\,\text{mm} \tag{6.9}$$

$$a_0 = 0.5 \cdot \ln\left(\frac{t}{t_0}\right) \text{ for } t \geq 15\,\text{mm} \tag{6.10}$$

$t_0 = 1$ mm is the reference thickness. The surface length of the crack is given by the crack shape, a_0/c_0, in function of the detail type:

- Non-welded details and longitudinal attachments: $a_0/c_0 = 0.40$
- Details with transverse welds : $a_0/c_0 = 0.15$

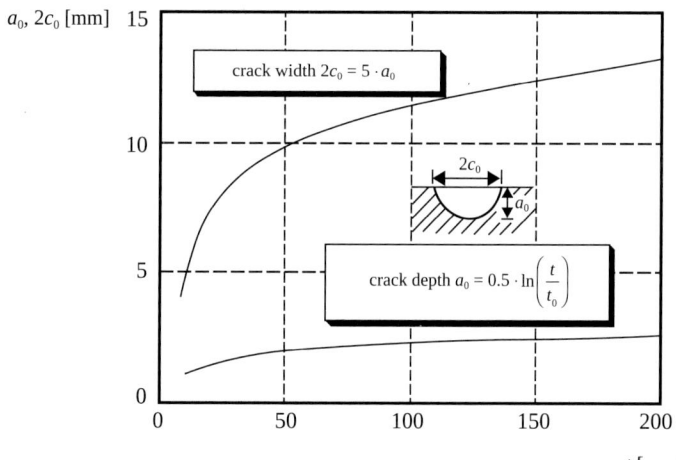

Figure 6.8 – Flaw in a plate (non-welded details), modelled as an initial crack with dimensions a_0 and c_0 (JRC, 2008)

Under service conditions, the initial crack may grow due to fatigue loading to a size a_d until it is detected during an inspection or failure occurs

6.4 Fracture Concept in EN 1993-1-10

before the next inspection, see Figure 6.9. It is this last case that has to be avoided by proper choice of material, with a material tough enough to tolerate such a crack, using the safety verification presented in equation (6.3) or equation (6.7). The other possible failure mode, by yielding of the remaining section is obviously less critical but can also be considered using the R6 FAD diagram approach (BS7910, 1999).

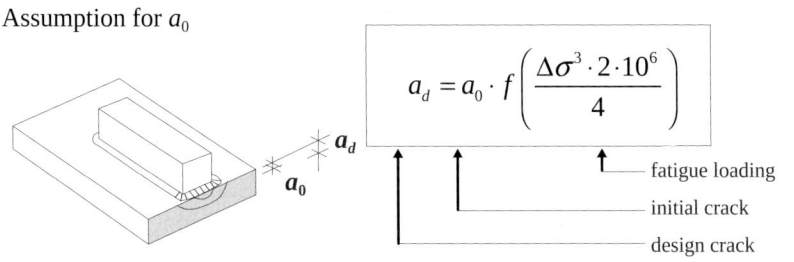

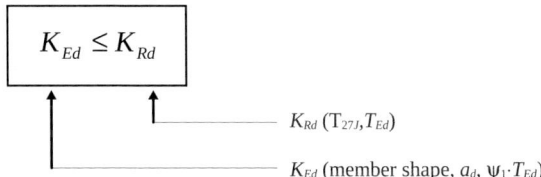

Figure 6.9 – Schematic of the method and main parameters in the fracture mechanics safety verification (Schmackpfeffer et al, 2005)

Using again fracture mechanics and making assumptions about fatigue loads, computations of the crack growth between two inspections can be made to get the design crack size a_d. Many computations were made during the validation of EN 1993-1-10. A conservative estimate for the design crack size a_d, in mm, is given in equation (6.11).

$$a_d = \left(2 \cdot 10^{-6} \cdot t^3 + 6 \cdot 10^{-4} \cdot t^2 + 0.134 \cdot t + 0.635\right) \left(\frac{\Delta \sigma_C}{\Delta \sigma_{ref}}\right)^3 \quad (6.11)$$

where $\Delta\sigma_{ref}$ = 56 N/mm² is the reference stress range and t the plate thickness (in which the fatigue crack grows) (JRC, 2008). The crack shape evolves with crack depth and the design crack shape a_d/c_d may be taken as

6. BRITTLE FRACTURE

0.40 for all details (i.e. it evolves from 0.15 to 0.40 in the case of details with transverse welds). Explanations about the stresses and stress ranges acting on the structural member are given in the next section.

6.4.4 Design value of the action effect stresses

The design value of the action effect stresses σ_{Ed} is expressed as a portion of the yield strength. The combination of actions to consider here, a frequent load combination, is the accidental one with the low temperature as the accidental action, see EN 1990:

$$E_d = E\left\{G_k; T_k; \psi_1 \cdot Q_{K,1}; \psi_{2,i} \cdot Q_{K,i}\right\} \tag{6.12}$$

where
- G_k characteristic nominal value of the permanent actions effects,
- $Q_{K,1}$ characteristic value of dominant variable load, usually the traffic load,
- $Q_{K,i}$ characteristic value of accompanying variable loads,
- ψ_1 combination factor for frequent loads,
- $\psi_{2,i}$ combination factor for quasi-permanent loads,
- T_k characteristic value of the lowest service temperature.

The action effect stresses σ_{Ed} from the addition of the following two sources:

1) applied stress resulting from the frequent load combination as explained above, and expressed in equation (6.13).

$$\sigma_{app} = \sigma\left(\sum G_k + \psi_1 Q_{K,1} + \sum \psi_{2,i} Q_{K,i}\right) \tag{6.13}$$

where
- G_k characteristic nominal value of the permanent actions effects,
- $Q_{K,1}$ characteristic value of dominant variable load, usually the traffic load,
- $Q_{K,i}$ characteristic value of accompanying variable loads,
- ψ_1 combination factor for frequent loads, see EN 1990,
- $\psi_{2,i}$ combination factor for quasi-permanent loads, see EN 1990.

6.4 Fracture Concept in EN 1993-1-10

2) residual stress σ_S in the structural member resulting from the fabrication process, in particular from welding.

Regarding residual stresses, the value of local residual stresses, for example at the hot spot of a detail, varies a lot through the thickness and thus some kind of average has to be taken. Indeed one cannot in this case take a conservative value of the residual stress field equal to the yield stress as the structural members would not be able to carry any external loads. Thus, local residual stresses are implicitly included in the residual stress value in EN 1993-1-10. It was found that a value of $\sigma_S = 100$ MPa could be used.

Regarding crack growth, it can be shown that the evolution of crack size and damage sum are affine curves, they both are exponential curves and function of stress range to the power m and number of cycles n_i. One shall first express the damage sum. For a stress range spectrum ($\Delta\sigma_i - n_i$), it can be expressed using the S-N curve equation (1.2) and the damage equation (1.4) as given in equation (6.14).

$$\sum_{i=1}^{n_{tot}} \frac{n_i}{C\Delta\sigma_i^{-m}} = \frac{\sum n_i}{C\Delta\sigma_E^{-m}} \leq 1.0 \qquad (6.14)$$

The above can also be expressed in function of the detail category and the corresponding S-N curve as given in equation (6.15). The slope coefficient is either $m = 3$ or, for bridges, $m = 5$ since the majority of the stress ranges levels are assumed to be between $\Delta\sigma_D$ and $\Delta\sigma_L$.

$$\frac{\sum \left(n_i \cdot \Delta\sigma_E^m\right)}{2 \cdot 10^6 \cdot \Delta\sigma_C^m} \leq 1.0 \qquad (6.15)$$

In EN 1993-1-10, to cover all relevant fatigue classes in EN 1993-1-9, the worst case is considered and thus it is assumed any structural detail is subjected to the maximum possible load range a detail can bear with a survival probability of 95 %, that is its detail category $\Delta\sigma_C$. Assuming further that the damage sum reaches unity at the end of the complete working life, one can see that, from equation (6.15), damage sum evolution over the working life (e.g. number of cycles n_i) can be simplified and represented using the detail category as an equivalent fatigue loading, i.e. $\sum \left(n_i \cdot \Delta\sigma_i^m\right) = 2 \cdot 10^6 \cdot \Delta\sigma_C^m$.

6. BRITTLE FRACTURE

In EN 1993-1-10, the so-called "safe service period" is defined as ¼ of the limiting damage value (e.g. value 1.0), which means 3 inspections during the service life, see sub-chapter 5.2 and section 5.3.3. Note here that the corresponding reliability index varies in function of the consequence of failure. Assuming a linear damage evolution, which is a gross assumption but conservative for the start of the damaging process, ¼ of the limiting damage value can be assumed to be reached after applying 500 000 cycles at a constant stress range equal to $\Delta\sigma_C$.

Regarding crack growth from initial flaws a_0 up to their design value a_d, the same assumptions as for damage sum evolution are made about the fatigue loading, i.e. crack growth is computed by applying 500 000 cycles at a constant stress range $\Delta\sigma_C$ to initial flaws a_0. Many details from EN 1993-1-9 were analysed, experimentally and numerically, to check with more refined calculations the assumptions on damage sum and crack growth evolutions. As a result, the conservative expression for a_d given in equation (6.11) was proposed. Note that EN 1993-1-10 background (JRC, 2008) explains that after the "safe service period", an inspection of the structure is carried out. The outcome of it can lead to one of the two following situations:

- if no damages are detected, the presence of undetected initial cracks a_0 (same sizes as after fabrication) may be assumed and a new "safe service period" may start,
- if damages are detected, relevant measures for repair or retrofitting should be taken before a new "safe service period" may start.

6.5 STANDARDISATION OF CHOICE OF MATERIAL: MAXIMUM ALLOWABLE THICKNESSES

The previous section showed the complexity of the computations associated with safety verification according to equation (6.3) or equation (6.7). Thus, a simplified procedure for the choice of material is necessary. From the explanations given in the previous sections, one can see that the thickness of the material or structural member is an important parameter. It can be found in equation (6.4), which contains the thickness expressed as b_{eff} and the initial as well as the design crack sizes depend on thickness, equations (6.9) to (6.11). Thus, since the structural member thickness influences both fatigue strength and

6.5 STANDARDISATION OF CHOICE OF MATERIAL: MAXIMUM ALLOWABLE THICKNESSES

brittle fracture, it was decided to develop tables in function of the thickness t of the member to be designed against brittle fracture.

The verification relationship is the same as given in equation (6.7), with equation (6.8), but it is simplified since several temperature shifts input values are directly included in the tables. The resulting verification is given in expression (6.16) below.

$$T_{Ed} = T_{min,d} + T_r + \Delta T_{\dot{\varepsilon}} + \Delta T_{\varepsilon pl} \geq T_{Rd} \qquad (6.16)$$

The permissible plate thicknesses of structural members with the most common structural details are given in function of:

- the steel grades and their toughness properties,
- the reference temperatures T_{Ed}, which usually is simply the lowest air temperature $T_{min,d}$ with a specified return period, see EN 1991-1-5,
- stress levels σ_{Ed}, three in total, which implicitly include the average residual stress level.

Three stress levels values σ_{Ed} from "frequent loads" have been fixed in the development of the tables in function of the yield stress, $f_y(t)$, namely:

- $\sigma_{Ed} = 0.25 f_y(t)$ This value corresponds to low stressed structural members.
- $\sigma_{Ed} = 0.50 f_y(t)$ This values corresponds to a medium case of loaded structural members.
- $\sigma_{Ed} = 0.75 f_y(t)$ This value corresponds to the maximum possible "frequent stress", where for the ultimate limit state verification yielding of the extreme fiber of the elastic cross section has been assumed.

For intermediate σ_{Ed} values, linear interpolation can be used. The yield stress value $f_y(t)$ may be determined either from equation (6.17) or taken as R_{eH}-values from the relevant steel product standards.

$$f_y(t) = f_y - 0.25 \cdot \left(\frac{t}{t_0}\right) \qquad (6.17)$$

where t is the thickness of the plate in mm and $t_0 = 1$ mm.

The stress level considered in this selection is corresponding to something similar to an "accidental load combination", owing to the fact that it

6. BRITTLE FRACTURE

occurs with the assumption of having simultaneously the lowest temperature, the presence of a crack, the lowest admissible material properties and the frequent loads. Once the material is properly selected using these tables, it can be assumed that fatigue cracking can occur without resulting in a brittle fracture and thus fatigue verification can be undertaken using either a damage tolerant approach or a safe life one, see sub-chapter 5.2.

Annex C contains reproductions of the tables from EN 1993-1-10, respectively EN 1993-1-12. Table C.1 in Annex C gives the permissible plate thicknesses for steels S235 to S690 (according to EN 10025). Table C.2 gives the permissible plate thicknesses for S500 to S700 (according to EN 10025-6 and EN 10149). Note that these tables are not applicable to other steel products, in particular hollow sections according to EN 10210 and EN 10219. Particular care should be taken with cold-formed sections subsequently welded, because significant changes occur in material properties, including toughness. An extension of EN 1993-1-10 to these products is currently under study (Feldmann *et al*, 2010).

Example 6.1: Application to steel and concrete composite road bridge, validation of the steel quality of the upper flanges.

The thicker plates are the one used for the upper flanges, see Figure 1.15. Their dimensions are comprised between 1500×50 and 1500×100 mm. They are made out structural steel grade S355 and chosen in quality NL, see Table 1.7.

The quality is determined using the table from EN 1993-1-10. Tables can be used since butt welds between flanges details are covered by EN 1993-1-9. The reference temperature T_{Ed} is thus determined as follows (equation (6.8)):

Minimum air temperature $T_{min.d} = -20$ °C (assuming the bridge is in Paris region, the value found in NAD of EN1991-1-5)

Radiation loss of member, $\Delta T_r = -5$ °C (according to EN 1991-1-5 for a steel and concrete composite bridge)

The influence of shape and dimensions of the member, imperfection from crack, and stress, $\Delta T_\sigma = 0$ °C

The additive safety element, $\Delta T_R = 0$ °C
Note that the above influences are already included when using tabulated values.

6.5 STANDARDISATION OF CHOICE OF MATERIAL: MAXIMUM ALLOWABLE THICKNESSES

Additionnal influences are the influence of strain rate, with assumption from project specification that there are no high strain rates effects ($\dot{\varepsilon} = 0.0001$ s^{-1}), thus $\Delta T_{\dot{\varepsilon}} = 0$ °C

And finally, the influence from cold forming, irrelevant here (DCF = 0), $\Delta T_{\varepsilon pl} = 0$ °C

The resulting value for the reference temperature is $T_{Ed} = -25$ °C

The relevant stress σ_{Ed} is calculated with the accidental combination of actions considering the low temperature as the accidental load case :

$$A\left[T_{Ed}\right] + \sum G_k + \psi_1 Q_{k,1} + \psi_2 Q_{k,2}$$

where $\psi_1 Q_{k,1}$ is the frequent value of the traffic loads and $\psi_2 Q_{k,2}$ is the quasi-permanent value for the other eventual variable loads.

Figure 6.10 gives the variation of the direct stresses in the upper flange of the box girder for this combination of actions.

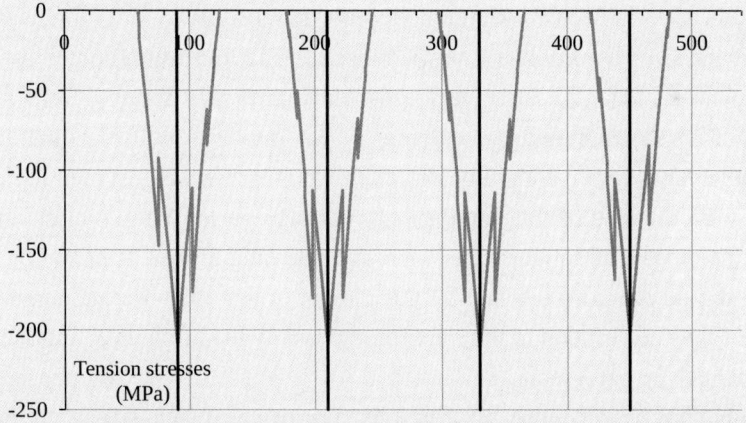

Figure 6.10 – Tensile stresses in the upper flange for the determination of the steel quality

The maximum design stress σ_{Ed} is equal to 211.9 N/mm² on internal support P3 where the flange thickness is 100 mm.

$$\sigma_{Ed} = \frac{211.9}{330} f_y(t) = 0.64 f_y(t)$$

Note: the yield stress must be determined using equation (6.17). It results in a higher value than the one given in the production standard for 100 mm thick plates (315 N/mm²).

6. BRITTLE FRACTURE

The use of table 6.11 (table 2.1 in EN 1993-1-10) requires interpolation. For a S555NL steel, the following permissible thickness values are extracted from the table:

$T_{Ed} = -30\,°C, \sigma_{Ed} = 0.50 f_y(t): t \leq 110$ mm

$T_{Ed} = -30\,°C, \sigma_{Ed} = 0.75 f_y(t): t \leq 75$ mm

$T_{Ed} = -20\,°C, \sigma_{Ed} = 0.50 f_y(t): t \leq 135$ mm

$T_{Ed} = -20\,°C, \sigma_{Ed} = 0.75 f_y(t): t \leq 90$ mm

Thus, for $\sigma_{Ed} = 0.64 f_y(t)$ and $T_{Ed} = -25\,°C$, the permissible thickness is 100.2 mm. Thus, for $t_{max} = 100$ mm, the steel quality NL is appropriate. Note: it means that the minimum toughness requirement is $T_{KV27} = -50\,°C$.

Tables reproduced in Annex C were developed with the assumption that the most onerous case of structures susceptible to fatigue must be included. That is a case where the design crack, with dimensions a_d, $2c_d$, does not only cover crack sizes overlooked in inspections after fabrication (denoted as initial cracks with a_0, $2c_0$), but also cracks that have grown in service, from fatigue actions from the moment the structure is put into use until the moment the cracks that have grown should be detected. Crack growth is assumed to depend not only on the size of the initial crack, but also on the fatigue class and the fatigue loading. The fatigue resistance and the fatigue load applied for crack growth should cover all relevant fatigue classes in EN 1993-1-9 and are defined such that they correspond to the maximum possible load in fatigue assessments, as explained in paragraphs 6.4.3 and 6.4.4. Figure 6.11 shows the limiting curve obtained using these assumptions, i.e. design crack size according to equation (6.11) and the case of a design stress level $\sigma_{Ed} = 0.75 f_y$, expressed applied stress temperature shifts ΔT_σ.

In order to validate the procedure and its conservatism, crack growth computations were made using boundary elements models (JRC, 2008) for various typical details and practical design situations (attachment on the flange of a girder, butt-joint, etc.). The results of those computations are shown in Figure 6.11. Safe-sided computation results must lie below the limiting curve. As can be seen, all crack growth computations fall below the limit curve and therefore the assumptions made for building Tables in Annex C are safe.

6.5 STANDARDISATION OF CHOICE OF MATERIAL: MAXIMUM ALLOWABLE THICKNESSES

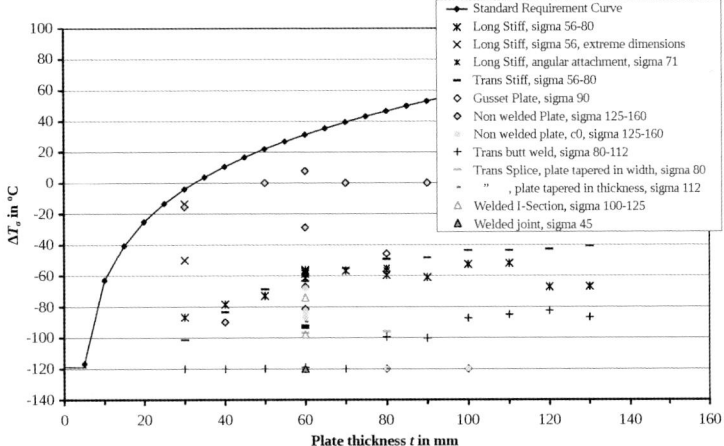

Figure 6.11 – Comparison between results from crack growth computations for various typical details in steel S355 and ΔT_σ limiting curve obtained with $\sigma_{Ed} = 0.75 f_y$
(JRC, 2008)

In order to get visualise the influence of the different parameters on the maximum permissible thickness of an element, the values in the tables where used to produce Figure 6.12 and Figure 6.13. Figure 6.12 shows, for steel grade S235, the influence of the action effects (stress level E_d) and of the subgrade. Finally, Figure 6.13 shows, the influence of the steel grade, for medium action effects (stress level $E_d = 0.50 f_y$) and only a few selected subgrades, on maximum permissible thickness of an element. From these figures, for example one can see that a 100 mm thick plate in S235 can be used in almost any situation, while it cannot be used at all for a S690 steel (max. 95 mm).

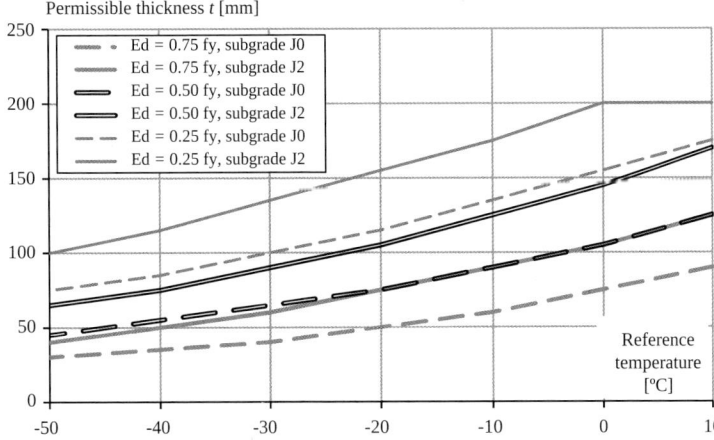

Figure 6.12 – Maximum permissible thickness of an element, influence of stress level and of the subgrade, for steel grade S235

6. BRITTLE FRACTURE

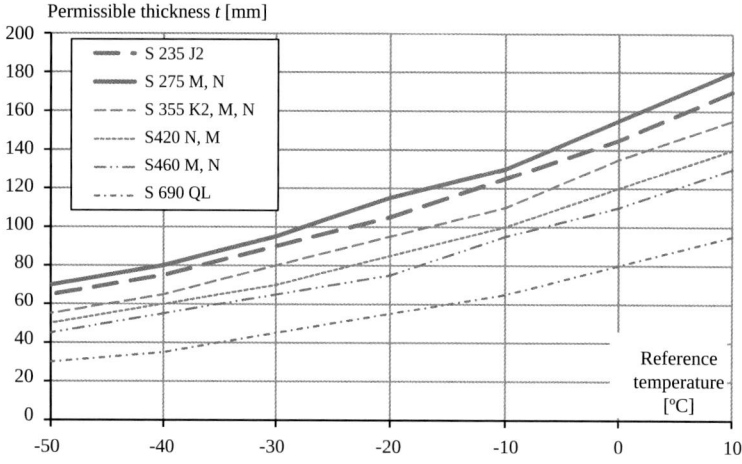

Figure 6.13 – Maximum permissible thickness of an element, influence of steel grade, from S2365 to S690, for selected subgrades

Example 6.2: Application to runway beam of crane, choice of the steel quality of the runway beam for dynamic load effects.

The runway beam is a rolled I-beam HEA 280, with flange thicknesses equal to 13 mm. The rail is a KSN 50×30 mm. Both are made out structural steel grade S235. The quality has not been yet defined.

The quality is determined using the table from EN 1993-1-10. Tables can be used as the different details of the runway beam are standard and covered by EN 1993-1-9. The reference temperature T_{Ed} is determined as follows:

- Minimum air temperature $T_{min,d} = 5\ °C$ (inside)
- Radiation loss of member, $\Delta T_r = 0\ °C$ (stable minimum temperature)
- Influence of strain rate, with assumption from project specification that a maximum strain rate $\dot{\varepsilon} = 0.005\,s^{-1}$ can occur (choc [JRC, 2008])

$$\Delta T_{\dot{\varepsilon}} = -\frac{1440 - f_y(t)}{550}\left(\ln\frac{\dot{\varepsilon}}{\dot{\varepsilon}_0}\right)^{1.5} = -\frac{1440 - 227}{550}\left(\ln\frac{0.005}{0.0001}\right)^{1.5} = -17.0\ °C$$

With yield stress determined using equation (6.17), leading to $f_y(t) = 227\ N/mm^2$ for the maximum thickness, which is 30 mm.

- No influence from cold forming (DCF = 0), $\Delta T_{\varepsilon pl} = 0\ °C$

6.5 STANDARDISATION OF CHOICE OF MATERIAL: MAXIMUM ALLOWABLE THICKNESSES

Note that are already included when using tabulated values:
- The influence of shape and dimensions of the member, imperfection from crack, and stress, $\Delta T_\sigma = 0\ °C$
- The additive safety element, $\Delta T_R = 0\ °C$

The resulting value of the reference temperature is then, equation (6.8):

$$T_{Ed} = T_{min,d} + \Delta T_r + \Delta T_{\sigma.} + \Delta T_R + \Delta T_{\dot{\varepsilon}} + \Delta T_{\varepsilon pl} = -12.0\ °C$$

The relevant stress σ_{Ed} is calculated with the accidental load combination and $\psi_2 = 0.7$. The total design stress is computed from the ULS design which is made in (TGC11, 2006). That is

$$\sigma_{Ed} = 1.0 \cdot (2.1) + 0.7 \cdot (55.3 + 53.9) = 79\ N/mm^2$$

$$\sigma_{Ed} = \frac{79}{227} f_y(t) = 0.35 \cdot f_y(t)$$

The interpolation can then be made with EN 1993-1-10 table 2.1. to determine steel quality for the maximum thickness ($t = 30$ mm), and vice-versa the maximum permissible thickness for a steel quality.

Assuming the choice of subgrade JR, one gets by successive interpolation:

$$T_{Ed} = -12\ °C,\ \sigma_{Ed} = 0.25 f_y(t):\ t \leq 65 - (2/10)(65-55) = 63\ mm$$

$$T_{Ed} = -12\ °C,\ \sigma_{Ed} = 0.50 f_y(t):\ t \leq 100 - (2/10)(100-85) = 97\ mm$$

Thus, for $\sigma_{Ed} = 0.35 \cdot f_y(t)$, the permissible thickness is 83 mm and the steel quality JR is appropriate.

REFERENCES

Note: all European and international standards references have been regrouped into Annex A and are thus not repeated here.

Akesson B (2008). *Understanding bridge collapses*, CRC press, USA, ISBN 978-0-415-43623-6, 280 p.

ASM Handbook (1996). *Fatigue and fracture*, Volume 19, ASM international, Materials Park, OH, USA, ISBN 0-87170-385-8.

Bäckström M and Marquis G (2004). Interaction equations for multiaxial fatigue assessment of welded structures, Fatigue & Fracture of Engineering Materials & Structures, Blackwell Publishing Ltd., Vol. 27, Issue 11, pp. 991 – 1003.

Banz A, Nussbaumer A (2001). *Fiabilité des ponts métalliques vis-à-vis de l'état limite combiné fatigue et rupture*, Report ICOM n° 446, ICOM – Steel structures, EPFL, Lausanne.

Barsom JM (1975). Development of the AASHTO Fracture-Toughness requirements for bridges steels, Engineering Fracture Mechanics, 7, N° 3.

Barsom JW and Rolfe ST (2000). *Fracture and Fatigue Control in Structures*, 3rd edition, Publ. Butterworth-Heinemann.

Bathias C, Paris PC (2005). *Gigacycle Fatigue in Mechanical Practice*. Publisher: Marcel Dekker. New York. 295 p.

Beg D, Kuhlmann U, Davaine L, Braun B (2010). *Design of Plated Structures, Eurocode 3, Part 1-5: Design of plated structures*, ECCS Eurocode design manuals, Editor ECCS, Publisher Ernst & Sohn, Berlin.

Broek D (1986). *Elementary engineering fracture mechanics*, 4th revised edition, Kluwer Academic Publ..

Castillo E, Lopez-Aenlle M, Ramos A, Fernandez-Canteli A, Kieselbach R, Esslinger V (2006). *Specimen length effect on parameter estimation in modeling fatigue strength by Weibull distribution*, International Journal of Fatigue, Elsevier science, Vol. 28, pp. 1047–1058.

REFERENCES

Cluni F, Gusella V and Ubertini F (2007). *A parametric investigation of wind-induced cable fatigue*, Engineering Structures, Elsevier Science, Vol. 29, N° 11, pp.3094-3105.

CIDECT (2001). CIDECT recommendations, 2001, Zhao, X. L., Herion, S., Packer, J. A. and al., *Design guide for circular and rectangular hollow section joints under fatigue loading.* CIDECT, Comité International pour le développement et l'étude de la construction tubulaire, guide no 8, TüV-Verlag Rheinland, Köln (downloadable from : http://www.cidect.org/en/Publications/).

COMBRI (2007). *Competitive steel and composite bridges by improved steel plated structures.* Final report, RFCS European Research Project RFS-CR-03-018, Brussels.

COMBRI+ (2008). *Design Manual - Part I: Application of Eurocode rules.* Final report, RFCS European Research Project RFS2-CT-00-031, Brussels.

Calgaro JA, Tschumi M, Shetty N, Gulvanessian H (2009). *Designers' Guide to Eurocode 1: Actions on Bridges*: EN 1991-2, EN 1991-1-1, -1-3 to 1-7 and EN 1990 Annex A2, Thomas Telford Ltd, 256 p.

Davaine L, Raoul J, Imberty F (2007). *Eurocodes 3 and 4, Application to steel and concrete composite road bridges,* Guidance book, Published by SETRA's Collection, Paris.

Koning CHM, Wardenier J, Dutta D (1992). Fatigue behaviour of multiplanar welded hollow section joints and reinforcement measures for repair. TNO-Bouw Report, No. BI-92-0005 / 21.4.6394.

DIN-Fachbericht 101 (2003). Einwirkungen auf Brücken, *Deutsches Institut für Normung e.V.*

DNV (2010). DNV-RP-C203, fatigue design of offshore steel structures, recommended practice, *Det Norske Veritas*, Norway. www.dnv.com:6389/dynaweb/offshore/rp-c203

DNV (2010.2). DNV-OS-C101, design of offshore steel structures, general (LRFD method), offshore standard, *Det Norske Veritas*, Norway, 2005. webshop.dnv.com

Dong P (2001). Structural stress definition and numerical implementation for fatigue analysis of welded joints, Intl. Journal of Fatigue, Elsevier Science, Vol. 23, pp. 865-876.

Dong P, Hong JK (2006). A Robust Structural Stress Parameter for Evaluation of Multiaxial Fatigue of Weldments, Journal of ASTM International, Vol. 3, Issue 7.

ECCS (1985). *Recommendations for the Fatigue Design of Steel Structures*, P043, European Convention for Constructional Steelwork, Brussels.

ECCS (1992). *European Recommendations for Aluminium Alloy Structures Fatigue Design*, P068, European Convention for Constructional Steelwork, Brussels.

ECCS (2000). *Good design practice – A guideline for fatigue design*, P105, European Convention for Constructional Steelwork, Brussels.

English CA and Hyde JM (2007). Radiation Embrittlement of Reactor Pressure Vessel Steels, comprehensive structural integrity, Vol. 6, Elsevier Science, pp. 351-398.

Euler M, Kuhlmann U (2009). Beitrag zur Ermüdungsfestigkeit von Schienenschweißnähten bei Kranbahnen, Ernst & Sohn, Berlin, Stahlbau 78, no. 3.

ESDEP (1995). European Steel Design Education Program, the ESDEP Society, *the Steel Construction Institute*, England. www.esdep.org

Feldmann M, Pinger T, Sedlacek G, Tschickardt D (2008). The new DASt-guideline to prevent cracking during hot-clip galvanizing (in german: Die neue DASt-Richtlinie zur Vermeidung von Rissbildung beim Feuerverzinken), Ernst & Sohn, Stahlbau, vol. 77, no 10, pp. 734-742.

Feldmann M, Eicher B, Kühn B, Sedlacek G (2010). *Choice of material for hollow section structures*, CEN TC 250 SC3, Working doc. N1729.

Fisher JW, Nussbaumer A, Keating PB, Yen BT (1993). *Resistance of welded details under variable amplitude long-life fatigue testing*, National Cooperative Highway Research Program (NCHRP) Report 354, Highway Research Board, Washington DC.

FKM (2006). FKM Recommendation, *Fracture mechanics proof of strength for engineering components* (original in german, Bruchmechanischer Festigkeitnachweis für Maschinbauteile, www.fkm-guideline.de), Forschungskuratorium Maschinenbau (FKM), 3rd edition.

References

FOSTA (2010). Wirtschaftliches Bauen von Straßen- und Eisenbahnbrücken aus Stahlhohlprofilen (Economic use of structural hollow sections for highway and railway bridges), Final report, Project FOSTA P591, Forschungsvereinigung Stahlanwendung e.V., Düsseldorf, Germany, 134 p.

Gough HJ and Pollard HV (1935). The strength of metals under combined alternating stresses, Proc. Inst. Mech Engrs, Vol.131, pp.3-103.

Gough HJ, Pollard HY (1951). *Some experiments of the resistance of metals to fatigue under combined stresses*, Min. of Supply, Aero Res. Council, RSM 2522, Part I.

Jean-Paul Gourmelon (2007). Exécution des structures métalliques, la nouvelle norme européenne EN 1090 (Execution of steel structures, the European standard EN 1090), bulletin Ouvrages Métalliques, OTUA, n° 5, pp. 8-26.

Gurney TR (1979). *The influence of thickness on the fatigue strength of welded joints*, 2nd International Conference on the Behaviour of Offshore Structures, Vol. 1, Paper 41, pp. 523-534. BHRA Fluid Engineering, Cranfield, Bedford, England.

Haibach E (1970). Modified linear damage accumulation hypothesis accounting for a decreasing fatigue strength during increasing fatigue damage. Report TM Nr. 50. Darmstadt: Laboratorium für Betriebsfestigkeit, LBF; German.

Hendy C, Johnson R (2006). *Designer's guide to EN1994-2*, Published by Thomas Telford, ISBN: 0727731610, UK.

Hobbacher A (2003). *Comparison of methods of fatigue analysis at an example of cruciform fillet welded joints*. In Proc. Int. Conf. Metal Structures, Miskolc Hungary April 3-5 2003, K. Jarmi and J Farkas eds., Millpress Rotterdam 2003, ISBN 90 77017 75 5, pp. 11-18.

Hobbacher A, Kassner M (2010). On Relation between Fatigue Properties of Welded Joints and Quality Groups in ISO 5817 - An Example for Discussion, IIW-Doc. JWG-XIII-XV-212-10 / XIII-WG4-106-10, Draft for development.

IIW (2000). Zhao XL, Packer, JA. *Fatigue design procedure for welded hollow section joints*, Doc. XIII-1804-99, XV-1035-99, IIW, Cambridge, Abington, 2000.

IIW (2009). International Institute of Welding (IIW): *Recommendations for fatigue design of welded joints and components*. A. Hobbacher (ed.), Doc. XIII-1965-08, IIW commission XIII.

IIW (2009b). International Institute of Welding (IIW), Draft ISO/WD 15.3, *Static Design Procedure for Welded Hollow Section Joints*, Recommendations (IIW Doc. XV-1329-09, IIW Doc. XV-E-09-400).

JRC (2008). Joint Research Centre (JRC), Commentary and Worked Examples to EN 1993-1-10 "Material Toughness and Through Thickness Properties" and Other Toughness Oriented Rules in EN 1993, G. Sedlacek, M. Feldmann, B. Kühn, D. Tschickardt, S. Höhler, C. Müller, W. Hensen, N. Stranghöner, W. Dahl, P. Langenberg, S. Münstermann, J. Brozzetti, J. Raoul, R. Pope, F. Bijlaard, Editors: M. Géradin, A. Pinto and S. Dimova, Report EUR 23510 EN – Scientific and Technical Research series – ISSN 1018-5593, Luxembourg. http://eurocodes.jrc.ec.europa.eu/showpublication.php?id=134

JRC (2012). M. Feldmann *et al*, Choice of Steel Material to Avoid Brittle Fracture for Hollow Section Structures, JRC Report EUR 25400 EN, Editors: A. Pinto, H. Amorim-Varum and B. Acun, 2012.

Kammel Ch (2003). *Worked chimney example for ECCS-TC6-WG-C*, RWTH Aachen, Germany, (unpublished).

Karamanos SA, Romeijn A, Wardenier J (1997). *Stress concentrations and joint flexibility effects in multi-planar welded tubular connections for fatigue design*. Stevin Report 6-98-05, CIDECT Report 7R-17/98, Delft University of Technology, The Netherlands.

Kawecki J, Zuranski JA (2007). *Cross-wind vibrations of steel chimneys - a new case history*, Journal of Wind Engineerig and Industrial Aerodynamics, Vol. 95, pp. 1166 – 1175.

Kolstein MK (2007). Fatigue classification of welded joints in orthotropic steel bridge decks, Doctoral thesis, Delft Univ..

Kuhlmann U, Dürr A, Günther H-P (2003). *Kranbahnen und Betriebsfestigkeit*, In: Kuhlmann, U. (Hrsg.), Stahlbaukalender, Ernst & Sohn, pp. 375-496.

Kuhlmann U, Günther H-P (2002). Ermüdungsverhalten von Trägern mit schlanken Stegblechen im Stahl- und Verbundbrückenbau, In: Stahlbau 71, No 6, pp. 460-469.

References

Landes JD (2000). The contributions of George Irwin to elastic-plastic fracture mechanics development, ASTM special technical publication, n°1389, pp. 54-63

Lechner A, Taras A (2009). *A numerical study of the fatigue proneness of flange thickness transitions in welded bridge girders,* proceedings of the 3rd international conference Fatigue Design, CETIM, Senlis, France.

Leendertz JS (2008). Fatigue Behaviour of Closed Stiffener to Crossbeam Connections in Orthotropic Steel Bridge Decks, Doctoral thesis, Delft Univ..

Lotsberg I (2009). Stress concentration due to misalignment at butt welds in plated structures and at girth welds in tubulars, Int. journal of fatigue, vol. 31, pp. 1337-1345.

Maddah N (2013). Fatigue Life Assessment of Roadway Bridges with Actual Traffic Loads, EPFL thesis n° 5575, Lausanne.

Maddox SJ (1997). *Developments in fatigue design codes and fitness-for-service assessment methods,* Proceedings IIW International Conference on Performance of Dynamically Loaded Welded Structures., Welding Research Council, New York.

McDiarmid DL (1994). A shear stress based critical-plane criterion of multiaxial fatigue failure for design and life prediction, Fatigue Fract. Eng. Mater. Struct, Vol. 17, No 12, pp. 1475-1484.

Manteghi S, Maddox SJ (2004). Methods for fatigue life improvement of welded joints in medium and high strength steels. International Institute of Welding, Doc. XIII-2006-04.

MacDonald KA, Maddox SJ (2003). *New guidance for fatigue design of pipeline girth welds.* Engineering Failure Analysis, Elsevier, Vol. 10, pp. 177–197.

Miner M (1945). *Cumulative Damage in Fatigue.* Journal of Applied Mechanics, Vol. 12, No. 3, pp. A159-A164.

Murphy C, Langner C (1985). "Ultimate pipe strength under bending, collapse and fatigue", Proceedings of the 4th International Conference on Offshore Mechanics and Arctic Engineering *(OMAE),* Dallas.

NCHRP (2002). Dexter RJ, Ricker MJ. Fatigue-*Resistant Design of Cantilevered Signal, Sign, and Light Supports,* Report NCHRP 469, National Cooperative Highway Research Program, NATIONAL ACADEMY PRESS WASHINGTON, D.C., USA.

Niemann HJ, Peil U (2003). *Windlasten auf Bauwerke. Stahlbau-Kalender*, Ernst & Sohn Verlag, S. 673-748.

Niemi E; Fricke W, Maddox SJ (2006). Fatigue Analysis of Welded Components - Designer's Guide to the Structural Hot-Spot Stress Approach, Woodhead Publ., Cambridge.

Nussbaumer A (2006). *European standard for fatigue design of steel structures and perspectives*, First International Conference on Fatigue and fracture in the Infrastructure, Philadelphia, USA.

Palmgren A (1923). *Die lebensdauer von kugellagern*, Z Ver Deut Ing, vol. 68, pp. 339–341.

Pargeter R (2003). *Liquid metal penetration during hot dip galvanizing*, TWI. http://www.twi.co.uk/j32k/unprotected/pdfs/ksrjp003.pdf

Petersen Chr (2000). *Dynamik der Baukonstruktionen*, Vieweg-Verlag, Germany.

Radaj D (2003). Ermüdungsfestigkeit: Grundlagen für Leichtbau, Maschine- und Stahlbau, 2nd Edition, Springer-Verlag, Berlin Heidelberg.

Romeijn A (1994). *Stress and strain concentration factors of welded multiplanar tubular joints*, Doctoral Thesis, Delft Univ. of Technology, Delft Univ. Press, The Netherlands.

Romeijn A, Karamanos SA, Wardenier J (1997). *Effects of joint flexibility on the fatigue design of welded tubular lattice structures*. 7th International Offshore and Polar Engineering Conference, Honolulu.

Ros M, Eichinger A (1950). *Die Bruchgefahr fester Körper bei wiederholter Beanspruchung d.h. Ermüdung*, EMPA-Bericht Nr. 173, Zurich, 167 pages (http://library.eawag-empa.ch/empa_berichte/EMPA_Bericht_173.pdf).

Sanz G, (1980). Essai de mise au point d'une méthode quantitative de choix des qualités d'aciers vis-à-vis du risque de rupture fragile, Revue de Métallurgie, CIT, pp. 621-642.

Schaumann P, Seidel M (2001). Ermittlung der Ermüdungsbeanspruchung von Schrauben exzentrisch belasteter Flanschverbindungen, Stahlbau 70, Heft 7, S. 474-486.

Schumacher A, Blanc A (1999). *Stress measurements and fatigue analysis on the new bridge at Aarwangen*, Report ICOM 386, ICOM-steel structures, EPFL, Lausanne.

References

Schumacher A (2003). *Fatigue behavior of welded circular hollow section joints*, PhD thesis no 2727, Ecole Polytechnique Fédérale de Lausanne (EPFL), Switzerland.

Schmackpfeffer H, Hensen W, Sedlacek G, and Müller Chr (2005). *Advantages of Eurocode 3 for the calculation of steel bridges*, 6th Japanese - German Bridge Symposium, Munich.

SED (2004). *Use and application of high performance steels (HPS) for steel structures*, Structural Engineering Document n° 8, editor: Hans-Peter Günther, IABSE, Zurich.

SED (2009). *Design for robustness*, structural Engineering Document n° 11, authors: Franz Knoll and Thomas Vogel, IABSE, Zurich.

Sedlacek G, Jaquemoud J (1984). *Herleitung eines Lastmodells für den Betriebsfestigkeitsnachweis von Straßenbrücken*, Forschung Straßenbau und Straßenverkehrstechnik, Heft 430.

Sedlacek G, Müller Ch (2000). Die Neuordnung des Eurocode 3 für die EN-Fassung und der neue Teil 1.9 – Ermüdung, Stahlbau 69, Heft 4, S. 228 – 235.

Sedlacek G, Kühn B, Höhler S, Stranghöner N, Langenberg P, Müller Chr (2002). *The application of fracture mechanics in steel construction*, Conference: la filière acier dans la construction – 30 ans d´innovation, Paris, ENCP.

Sedlacek G (2003). Leitfaden zum DIN-Fachbericht 103 „Stahlbrücken", Chapter II-9 Werkstoffermüdung, pp. 107-129.

Sedlacek G, Müller C, Nussbaumer A (2004). Chap. 4.2, Toughness requirements in structural applications, In: *Use and application of high performance steels (HPS) for steel structures*, Structural Engineering Document n° 8, editor: Hans-Peter Günther, IABSE, Zurich.

Schijve J (2001). *Fatigue of structures and materials*, Kluwer Academic Publ., The Netherlands (revised and reedited in 2009).

Schmidt H, Jakubowski A (2001). Zum Tragverhalten biegebeanspruchter vorgespannter L-Ringflanschstösse unter wiederholter Belastung, Festschrift zu Ehren von Prof. Dr. –Ing. G. Valtinat, Hamburg.

SETRA (2007). Davaine L, Imberty F, Raoul J. *Eurocode 3 and 4: application to steel-concrete composite road bridges*, Guidance book, SETRA, (downloadable from : http://www.jrc.eurocode.com).

Simiu E, Scanlan RH (1986). *Wind effects on structures, second edition*, John Wiley and Sons, New York.

Simões da Silva L, Simões R and Gervasio H (2010). *Design of steel structures, Eurocode 3 Part 1-1: General rules and rules for buildings*, ECCS Eurocode design manuals, Editior ECCS, Publisher Ernst & Sohn, Berlin.

Sonsino CM (2009). Multiaxial fatigue assessment of welded joints – Recommendations for design codes, International Journal of Fatigue, Vol. 31, pp. 173–187.

Stahlbaukalender (2006). Nussbaumer A, Gunther H-P. Ermüdung *Grundlagen und Erläuterung der neuen Ermüdungsnachweise nach Eurocode 3*, Stahlbaukalendar, Ed. U. Kuhlmann, Springer Verlag, Germany.

Stötzel J, Sedlacek G, Hobbacher A, Tschickardt D, Nussbaumer A (2007). *Background document to EN 1993-1-9*, Draft version (unpublished), RWTH Aachen, Germany.

Susmel L, Lazzarin P (2001). *A bi-parametric Wöhler curve for high cycle multiaxial fatigue asessment*, Fatigue Fract. Engng. Mater. Struct., Vol. 25, pp. 63-78.

Takena K, Miki C, Shimokawa H, Sakamoto K (1992). *Fatigue resistance of large-diameter cable for cable-stayed bridges*, Journal of Structural Engineering, ASCE, Vol. 118, No. 3, pp. 701-715.

TGC 10 (2006). Hirt, M., Bez, R., Nussbaumer, A. Construction Métallique – Notions fondamentales et méthodes de dimensionnement, PPUR Lausanne.

TGC 11 (2006). Hirt, M., Crisinel, M. Charpentes Métalliques – Conception et dimensionnement des halles et bâtiments, PPUR Lausanne, 2005.

VDI (2003). VDI-Richtlinie 2230, Systematische Berechnung hochbeanspruchter Schraubenverbindungen – Zylindrische Einschraubenverbindungen, Verein Deutscher Ingenieure, Berlin, reimpression with corrections.

Verwiebe C (2003). Wirbelerregte Querschwingungen von Industriebauwerken. In: Windwirkungen auf Bauwerke und deren Umgebung, Graubner (Hrsg.), WtG, ISBN 3-928909-07-X, S.33-40, Aachen.

References

Walbridge S (2005). *A probabilistic fatigue analysis of post-weld treated tubular bridge structures*. Ph.D. Thesis EPFL n°3330, Swiss Federal Institute of Technology Lausanne (EPFL), Switzerland.

Wallin K (1994). *Methodology for Selecting Charpy Toughness Criteria for Thin High Strength Steels, Part I, II and III*, Jernkontorets Forskning, No. 2013/89, TO40-05, -06, -31, VTT Manufacturing Technology, Finland.

Wardenier J (2011). Hollow Sections in Structural Applications, 2nd edition, CIDECT, (downloadable from: http://www.cidect.org/en/Publications/).

Wöhler A (1860). *Versuche uber die festigkeit der eisenbahnwagenachsen, zeitschrift fur bauwesen*, vol 10, 1860 (in German), with english summary in *engineering*, vol 4, 1867, p 160-161.

Zhao X-L, Haldar A, Breen F (1994). *Fatigue-Reliability Evaluation of Steel Bridges*. Journal of Structural Engineering, Vol. 120, No. 5, pp. 1608-1623, ASCE, May 1994.

Annex A

STANDARDS FOR STEEL CONSTRUCTION

AISC, (2005). *Steel Construction Manual, Thirteenth Edition*, American Institute of Steel Construction (AISC), Chicago, USA.

API, (2005). API RP 2A WSD – 22th Edition, with errata/supplements, *Recommended Practice for Planning, Designing and Constructing Fixed Offshore Platforms*, American Petroleum Institute (API), Washington DC, USA.

BS 5400, (1980). Steel, concrete and composite bridges. Code of practice for fatigue, *British Standard Institution*, London.

BS 7608, (1993). Fatigue of steel structures, *British Standard Institution*, London.

BS PD 7910, (1999). Assessment of flaws in welded structures, *British Standard Institution*, London.

DIN 15018, (1974). Cranes, Principles for steel structures; Stress analysis, *German Standards Organisation*, Berlin.

EN 1090-1, (2008). Execution of steel structures and aluminium structures – Part 1: Requirements for conformity assessment of structural components, *European Committee for Standardization*, Brussels.

EN 1090-2, (2008). Execution of steel structures and aluminium structures – Part 2: Technical requirements for steel structures, *European Committee for Standardization*, Brussels.

EN 1090-3, (2008). Execution of steel structures and aluminium structures – Part 3: Technical requirements for aluminium structures, *European Committee for Standardization*, Brussels.

EN 1990, (2002). Eurocode: Basis of design, *European Committee for Standardization*, Brussels.

A. Standards For Steel Construction

EN 1990:2002 A1, (2005). Eurocode: Basis of design - Amendment A1 to EN 1990:2002, *European Committee for Standardization*, Brussels.

EN 1991-1-1, (2002). Eurocode 1: Actions on structures - Part 1-1: General actions - Densities, self-weight, imposed loads for buildings, *European Committee for Standardization*, Brussels.

EN 1991-1-4, (2005). Eurocode 1: Actions on structures - Part 1-1: General actions – Wind actions, *European Committee for Standardization*, Brussels.

EN 1991-2, (2003). Eurocode 1: Actions on structures - Part 2: Traffic loads on bridges, *European Committee for Standardization*, Brussels.

EN 1991-3, (2006). Eurocode 1: Actions on structures – Part 3: Actions induced by cranes and machinery, *European Committee for Standardization*, Brussels.

EN 1991-4, (2006). Eurocode 1: Actions on structures – Part 4: Silos and tanks, *European Committee for Standardization*, Brussels.

EN 1993-1-1, (2005). Eurocode 3: Design of steel structures - Part 1-1: General rules and rules for buildings, *European Committee for Standardization*, Brussels.

EN 1993-1-2, (2005). Eurocode 3: Design of Steel Structures: Structural fire design, *European Committee for Standardization*, Brussels.

EN 1993-1-3, (2006). Eurocode 3: Design of Steel Structures: Cold-formed thin gauge members and sheeting, *European Committee for Standardization*, Brussels.

EN 1993-1-4, (2006). Eurocode 3: Design of Steel Structures: Stainless steels, *European Committee for Standardization*, Brussels.

EN 1993-1-5, (2006). Eurocode 3: Design of Steel Structures: Plated structural elements, *European Committee for Standardization*, Brussels.

EN 1993-1-6, (2007). Eurocode 3: Design of Steel Structures: Strength and stability of shell structures, *European Committee for Standardization*, Brussels.

EN 1993-1-7, (2007). Eurocode 3: Design of Steel Structures: Strength and stability of planar plated structures transversely loaded, *European Committee for Standardization*, Brussels.

EN 1993-1-8, (2005). Eurocode 3: Design of Steel Structures: Design of joints, *European Committee for Standardization*, Brussels.

A. STANDARS FOR STEEL CONSTRUCTION

EN 1993-1-9, (2005). Eurocode 3: Design of Steel Structures: Fatigue strength of steel structures, *European Committee for Standardization*, Brussels.

EN 1993-1-9/AC, (2009). Corrigendum to Eurocode 3: Design of Steel Structures: Fatigue strength of steel structures, *European Committee for Standardization*, Brussels.

EN 1993-1-10, (2005). Eurocode 3: Design of Steel Structures: Selection of steel for fracture toughness and through-thickness properties, *European Committee for Standardization*, Brussels.

EN 1993-1-11, (2006). Eurocode 3: Design of Steel Structures: Design of structures with tension components made of steel, *European Committee for Standardization*, Brussels.

EN 1993-1-12, (2007). Eurocode 3: Design of Steel Structures: Supplementary rules for high strength steel, *European Committee for Standardization*, Brussels.

EN 1993-2, (2006). Eurocode 3: Design of Steel Structures: Steel bridges, *European Committee for Standardization*, Brussels.

EN 1993-3-1, (2006). Eurocode 3: Design of Steel Structures: Towers and masts, *European Committee for Standardization*, Brussels.

EN 1993-3-2, (2006). Eurocode 3: Design of Steel Structures: chimneys, *European Committee for Standardization*, Brussels.

EN 1993-4-1, (2007). Eurocode 3: Design of Steel Structures: Silos, *European Committee for Standardization*, Brussels.

EN 1993-4-2, (2007). Eurocode 3: Design of Steel Structures: Tanks, *European Committee for Standardization*, Brussels.

EN 1993-4-3, (2007). EN 1993-4-3 - Eurocode 3: Design of Steel Structures: Pipelines, *European Committee for Standardization, Brussels*.

EN 1993-5, (2007). Eurocode 3: Design of Steel Structures: Piling, *European Committee for Standardization*, Brussels.

EN 1993-6, (2007). Eurocode 3: Design of Steel Structures: Crane supporting structures, *European Committee for Standardization*, Brussels.

EN 1994-1-1, (2004). Eurocode 4: Design of composite steel and concrete structures - Part 1-1: General rules and rules for buildings, *European Committee for Standardization*, Brussels.

A. Standards For Steel Construction

EN 1994-2, (2005). Eurocode 4: Design of composite steel and concrete structures - Part 2: General rules and rules for bridges, *European Committee for Standardization*, Brussels.

EN 10025, (2005). Hot rolled products of structural steels – Parts 1 to 6, *European Committee for Standardization*, Brussels.

EN 10045-1, (1990). Charpy impact test on metallic materials - Test method (V- and U-notches), *European Committee for Standardization*, Brussels.

EN 10088, (2005). Stainless steels. Technical delivery conditions, *European Committee for Standardization*, Brussels.

EN 10149, (1996). Specification for hot-rolled flat products made of high yield strength steels for cold forming, *European Committee for Standardization*, Brussels.

EN 10164, (1993). Steel products with improved deformation properties perpendicular to the surface of the product - Technical delivery conditions, *European Committee for Standardization*, Brussels.

EN 10210, (1994). Hot finished structural hollow sections of non-alloy and fine grain structural steels, *European Committee for Standardization*, Brussels.

EN 10219, (1997). Cold formed welded structural hollow sections of non-alloy and fine grain structural steels, *European Committee for Standardization*, Brussels.

EN 12062, (1997). Non-destructive testing of welds – General rules for metallic materials, *European Committee for Standardization*, Brussels.

EN 13001-1, (2004). Crane safety. General design. General principles and requirements, *European Committee for Standardization*, Brussels.

EN ISO 15611, (2003). Specification and qualification of welding procedures for metallic materials - Qualification based on previous welding experience, *International Organization for Standardization*, Geneva, Switzerland.

EN ISO 15612, (2004). Specification and qualification of welding procedures for metallic materials – Qualification by adoption of a standard welding procedure, *International Organization for Standardization*, Geneva, Switzerland.

EN ISO 15613, (2004). Specification and qualification of welding procedures for metallic materials - Qualification based on pre-production welding test, *International Organization for Standardization*, Geneva, Switzerland.

A. Standars for Steel Construction

EN ISO 15614, (2004). In 12 parts. Specification and qualification of welding procedures for metallic materials - Welding procedure test, *International Organization for Standardization*, Geneva, Switzerland.

ISO 12944, (1998). Paints and varnishes — Corrosion protection of steel structures by protective paint systems, *International Organization for Standardization*, Geneva, Switzerland.

ISO 12107, (2003). Metallic materials – Fatigue testing – Statistical planning and analysis of data, *International Organization for Standardization*, Geneva, Switzerland.

ISO 3834, (2005). Quality requirements for fusion welding of metallic materials , Parts 1 to 6, *International Organization for Standardization*, Geneva, Switzerland.

ISO 5817, (2005). Welding - Fusion-welded joints in steel, nickel, titanium and their alloys (beam welding excluded) - Quality levels for imperfections, *International Organization for Standardization*, Geneva, Switzerland.

ISO 9013, (2002). Thermal cutting - Classification of thermal cuts - Geometrical product specification and quality tolerances, *International Organization for Standardization*, Geneva, Switzerland.

ISO 15607, (2003). Specification and qualification of welding procedures for metallic materials - General rules, *International Organization for Standardization*, Geneva, Switzerland.

NORSOK, (2004). NORSOK standard N-004, Design of Steel structures, *Standards Norway*, rev.2.

Annex B

FATIGUE DETAIL TABLES WITH COMMENTARY

INTRODUCTION

This section reproduces the fatigue tables from EN 1993-1-9, as well as detail categories given in other Eurocode parts (cables, etc.). The tables include the corrections and modifications from the corrigendum issued in November 2008 (changes are highlighted with a gray background). In addition to the code, the tables contain an additional column with supplementary explanations and help for the engineer to classify properly fatigue details and compute correctly the stress range needed for the verification. For some details, suggestion from the authors about the required weld quality level (B or C) is given.

Then, the table for the use of the hot spot stress method (Annex B from EN 1993-1-9) is reproduced.

Finally, for orthotropic decks, the authors propose a new detail table in a attempt to clarify the EN 1993-1-9 tables 8.8 and 8.9. This table summarizes the detail information and includes notes on the potential modes of failure (crack location and consequences), important factors influencing the class of each detail type and some guidance on selection for design, including strength factor γ_{Mf}. This table is a combination and interpretation of propositions from different recent studies (Kolstein, 2007 and Leendertz, 2008).

B. Fatigue Detail Tables with Commentary

B.1 PLAIN MEMBERS AND MECHANICALLY FASTENED JOINTS (EN 1993-1-9, TABLE 8.1)

Detail category	Constructional detail	Description	Requirements	Commentary
	NOTE The fatigue strength curve associated with category 160 is the highest. No detail can reach a better fatigue strength at any number of cycles.			
160	① ② ③	Rolled or extruded products: 1) Plates and flats with as rolled edges; 2) Rolled sections with as rolled edges; 3) Seamless hollow sections, either rectangular or circular.	Details 1) to 3): Sharp edges, surface and rolling flaws to be improved by grinding until removed and smooth transition achieved.	Rolling flaws at edges are most detrimental. Those parallel to stresses do not need to be removed.
140	④	Sheared or gas cut plates: 4) Machine gas cut or sheared material with subsequent dressing;	4) All visible signs of edge discontinuities to be removed. The cut areas are to be machined or ground and all burrs to be removed.	For re-entrant stress concentration factors (SCF), use for example figure (3.22).
125	⑤	5) Material with machine gas cut edges having shallow and regular drag lines or manual gas cut material, subsequently dressed to remove all edge discontinuities. Machine gas cut with cut quality defined in EN 1090.	Any machinery scratches for example from grinding operations, can only be parallel to the stresses. Details 4) and 5): - Re-entrant corners to be improved by grinding (slope $\leq 1/4$) or evaluated using the appropriate stress concentration factors. - No repair by weld refill.	

	6) and 7) Rolled or extruded products as in details 1), 2), 3)	Details 6) and 7): $\Delta\tau$ calculated from: $\tau = \dfrac{V\,S(t)}{I\,t}$	This detail is seldom relevant.
100 $m = 5$ ⑥ ⑦			This downgrading is due to the negative influence from corrosion. Category 160 becomes 140, category 140 becomes 125, etc. For computing stresses and stress ranges, see section 3.7.3.
For detail 1 – 5 made of weathering steel use the next lower category.			
112 ⑧	8) Double covered symmetrical joint with preloaded high strength bolts.	8) $\Delta\sigma$ to be calculated on the gross cross section.	For bolted connections (Details 8) to 13)) in general: End distance: $e_1 \geq 1.5\,d$ Edge distance: $e_2 \geq 1.5\,d$ Spacing: $p_1 \geq 2.5\,d$ Spacing: $p_2 \geq 2.5\,d$ Detailing to EN 1993-1-8 Figure 3.1
	8) Double covered symmetrical joint with preloaded injection bolts.	8) … gross cross section.	
⑨	9) Double covered joint with fitted bolts.	9) … net cross section.	
	9) Double covered joint with non preloaded injection bolts.	9) … net cross section.	
90 ⑩	10) One sided connection with preloaded high strength bolts.	10) … gross cross section.	
	10) One sided connection with preloaded injection bolts.	10) … gross cross section.	
⑪	11) Structural element with holes subject to bending and axial forces	11) … net cross section.	
80 ⑫	12) One sided connection with fitted bolts.	12) … net cross section.	

B. Fatigue Detail Tables with Commentary

Cat.	Sketch	Description	Requirements
80	(8) g, h, g/h ≥ 2.5	8) Intermittent longitudinal fillet weld.	8) Δσ based on direct stress in flange. Welding quality level C could be specified.
71	(9)	9) Longitudinal butt weld, fillet weld or intermittent weld with a cope hole height not greater than 60 mm. For cope holes with a height > 60 mm see detail 1) in Table 8.4.	9) Δσ based on direct stress in flange. In case cope holes with a height > 60 mm, than the detail is equivalent to the end of a longitudinal attachment of length more than 100 mm; Thus use category 56.
125	(10)	10) Longitudinal butt weld, both sides ground flush parallel to load direction, 100% NDT.	Usually welding quality level B.
112		10) No grinding and no start/stop.	
90		10) With start/stop positions.	
140	(11)	11) Automatic or fully mechanized longitudinal seam weld without stop/start positions in hollow sections.	11) ~~Free from defects outside the tolerances of EN 1090.~~ Wall thickness t ≤ 12.5 mm.
125		11) Automatic or fully mechanized longitudinal seam weld without stop/start positions in hollow sections.	11) Wall thickness t > 12.5 mm. Usually welding quality level B.
90		11) with stop/start positions.	

For details 1 to 11 made with fully mechanised welding the categories for automatic welding apply.

B.3 TRANSVERSE BUTT WELDS (EN 1993-1-9, TABLE 8.3)

Detail category	Constructional detail	Description	Requirements	Commentary
112 size effect for $t > 25$ mm: $k_s = (25/t)^{0.2}$	(1) (2) (3) (4) slope 1/4	Without backing bar: 1) Transverse splices in plates, flats and rolled sections. 2) Flange and web splices in plate girders before assembly. 3) Full cross sections butt welds of rolled sections without cope holes. 4) Transverse splices in plates or flats tapered in width or in thickness, with a slope $\leq 1/4$.	- All welds ground flush to plate surface parallel to direction of the arrow. - Weld run-on and run-off pieces to be used and subsequently removed, plate edges to be ground flush in direction of stress. - Welded from both sides; checked by NDT. Detail 3) Applies only to joints of rolled sections, cut and rewelded.	For details 1) to 12) and 17) to 19) Welding quality level B. The classifications may be deemed to allow for the effects of any accidental axial or centreline misalignment up to the lesser of 0.15 times the thickness of the thinner part or 3 mm, provided that the root sides of joints with single-sided preparations i.e. single bevel-J, -U or -v forms are back-gouged over entire width (by analogy to BS 7608). For elements where additional bending is resisted by contiguous construction, e.g. beam flanges supported by webs (that is detail 2), wide plates supported by effectively continuous stiffeners, eccentricities due to axial misalignments in the thickness direction may be neglected, see section 3.4.2.
90 size effect for $t > 25$ mm: $k_s = (25/t)^{0.2}$	(5) (6) (7) slope 1/4 0.1b	5) Transverse splices in plates or flats. 6) Full cross sections butt welds of rolled sections without cope holes. 7) Transverse splices in plates or flats tapered in width or in thickness with a slope $\leq 1/4$. Translation of welds to be machined notch free.	- The height of the weld convexity to be not greater than 10% of the weld width, with smooth transition to the plate surface. - Weld run-on and run-off pieces to be used and subsequently removed, plate edges to be ground flush in direction of stress. - Welded from both sides; checked by NDT. Details 5 and 7: Welds made in flat position.	

B. Fatigue Detail Tables with Commentary

90 size effect for $t > 25$ mm: $k_s = (25/t)^{0.2}$	(8)	8) As detail 3) but with cope holes.	See comment for detail 1). All welds ground flush to plate surface parallel to direction of the arrow. Weld run-on and run-off pieces to be used and subsequently removed, plate edges to be ground flush in direction of stress. Welded from both sides; checked by NDT. - Rolled sections with the same dimensions without tolerance differences.
80 size effect for $t > 25$ mm: $k_s = (25/t)^{0.2}$	(9) (10) (11)	9) Transverse splices in welded plate girders without cope holes. 10) Full cross sections butt welds of rolled sections with cope holes. 11) Transverse splices in plates, flats, rolled sections or plate girders.	See comment for detail 1). 9) preferably with cope holes in web. The height of the weld convexity to be not greater than 20% of the weld width, with smooth transition to the plate surface. Weld not ground flush. Weld run-on and run-off pieces to be used and subsequently removed, plate edges to be ground flush in direction of stress. Welded from both sides; checked by NDT. Detail 10: The height of the weld convexity to be not greater than 10% of the weld width, with smooth transition to the plate surface.
63	(12)	12) Full cross sections butt welds of rolled sections without cope hole.	See comment for detail 1). Not recommended, high fatigue strength variability, no NDT check to insure quality, prefer detail 8). - Weld run-on and run-off pieces to be used and subsequently removed, plate edges to be ground flush in direction of stress. - Welded from both sides.

B.3 Transverse Butt Welds

Cat.	Formula	Sketch	Description	Comment
36		(13)	13) Butt welds made from one side only.	13) Without backing strip. Usually welding quality level C. If classified in cat. 71, then welding quality level B.
71	Size effect for $t > 25$ mm: $k_s = (25/t)^{0.2}$		13) Butt welds made from one side only when full penetration checked by appropriate NDT.	
71	Size effect for $t > 25$ mm: $k_s = (25/t)^{0.2}$	(14) (15)	With backing strip: 14) Transverse splice. 15) Transverse butt weld tapered in width or thickness with a slope $\leq 1/4$. Also valid for curved plates.	Details 14) and 15): Fillet welds attaching the backing strip to terminate ≥ 10 mm from the edges of the stressed plate. Tack welds inside the shape of butt welds. Usually welding quality level B. Insure very good alignment between curved plates.
50	Size effect for $t > 25$ mm: $k_s = (25/t)^{0.2}$	(16)	16) Transverse butt weld on a permanent backing strip tapered in width or thickness with a slope $\leq 1/4$. Also valid for curved plates.	16) Where backing strip fillet welds end < 10 mm from the plate edge, or if a good fit cannot be guaranteed.
71	Size effect for $t > 25$ mm and/or generalization for eccentricity: $k_s = \left(1 + \dfrac{6e}{t_1} \cdot \dfrac{t_1^{1.5}}{t_1^{1.5} + t_2^{1.5}}\right)\left(\dfrac{25}{t_1}\right)^{0.2}$ $t_2 \geq t_1$	(17)	17) Transverse butt weld, different thicknesses without transition, centrelines aligned.	See comment for detail 1)
40		(18) (19)	18) Transverse butt weld at intersecting flanges. 19) With transition radius according to Table 8.4, detail 4.	Details 18) and 19) The fatigue strength of the continous component has to be checked with Table 8.4, detail 4 or detail 5.
As detail 4 in Table 8.4				See comment for detail 1)

B. FATIGUE DETAIL TABLES WITH COMMENTARY

B.4 ATTACHMENTS AND STIFFENERS (EN 1993-1-9, TABLE 8.4)

Detail category	Constructional detail	Description	Requirements	Commentary
80	$L \leq 50$ mm	Longitudinal attachments: 1) The detail category varies according to the length of the attachment L.	The thickness of the attachment must be less than its height. If not, see Table 8.5, detail 6 or 7.	The stress concentration resulting from a longitudinal attachment is function of the relative proportions of its height, length and thickness. Thus, an attachment with its thickness greater than its height is in fact a coverplate (which length is always greater than 100 mm).
71	$50 < L \leq 80$ mm			
63	$80 < L \leq 100$ mm			
56	$L > 100$ mm			
71	$L > 100$ mm $\alpha < 45°$	2) Longitudinal attachments to plate or tube.		An as-welded attachment tapered with $\alpha \geq 45°$ does not change the stress concentration at the weld toe compared to detail 1).
80	$r > 150$ mm	3) Longitudinal fillet welded gusset with radius transition to plate or tube; end of fillet weld reinforced (full penetration); length of reinforced weld > T.	Details 3) and 4): Smooth transition radius r formed by initially machining or gas cutting the gusset plate before welding, then subsequently grinding the weld area parallel to the direction of the arrow so that the transverse weld toe is fully removed.	
90	$\dfrac{r}{\ell} \geq \dfrac{1}{3}$ or $r > 150$ mm	4) Gusset plate, welded to the edge of a plate or beam flange.		
71	$\dfrac{1}{6} \leq \dfrac{r}{\ell} \leq \dfrac{1}{3}$			
50	$\dfrac{r}{\ell} < \dfrac{1}{6}$			

L: attachment length as in detail 1, 2, or 3

B.4 ATTACHMENTS AND STIFFENERS

40		5) As welded, no radius transition.	
80	$\ell \leq 50$ mm	Transverse attachments: 6) Welded to plate. 7) Vertical stiffeners welded to a beam or plate girder. 8) Diaphragm of box girders welded to the flange or the web. May not be possible for small hollow sections! The values are also valid for ring stiffeners.	Details 6) and 7): Ends of welds to be carefully grounde to remove any undercut that may be present. 7) $\Delta\sigma$ to be calculated using principal stresses if the stiffener terminates in the web, see left side.	7) If the stiffener terminates near the flange, the shear stress range satisfies the rule less than 15% of the direct stress range and thus can be neglected. See section 5.4.7.2.
71	$50 < \ell \leq 80$ mm			
80		9) The effect of welded shear studs on base material.		Additional criteria for fatigue, extracts from EN 1994-2: - In the case of element under tension, the stud diameter shall be less than 1.5 of the thickness of the plate (6.6.5.7(3)). - An additional verification is required (6.8.1(3)): $\sigma_{Ed,max} \leq 0.75 P_{Rd}$.

B. Fatigue Detail Tables with Commentary

B.5 LOAD CARRYING WELDED JOINTS (EN 1993-1-9, TABLE 8.5)

Detail category	Constructional detail		Description	Requirements	Commentary
80	$\ell < 50$	all t	Cruciform and Tee joints: 1) Toe failure in full penetration butt welds and all partial penetration joints. 2) Toe failure from edge of attachment to plate, with stress peaks at weld ends due to local plate deformations. 2) Root failure in partial penetration Tee-butt joints or fillet welded joint and in Tee-butt weld, according to Figure 4.6 in EN 1993-1-8:2005.	1) Inspected and found free from discontinuities and misalignments outside the tolerances of EN 1090. 2) For computing $\Delta\sigma$, use modified nominal stress. 3) In partial penetration joints, two fatigue assessments are required. Firstly, root cracking evaluated according to stresses defined in section 5, using category 36* for $\Delta\tau_w$. Secondly, toe cracking is evaluated by determining $\Delta\sigma$ in the load-carrying plate. Details 1) and 2): The misalignment of the load-carrying plates should not exceed 15 % of the thickness of the intermediate plate.	1) and 2) usually welding quality level B 3) Welding quality level C could be specified. For computation of stresses in welds, see sections 3.3.4 and 3.7.4. For $\Delta\tau_w$, detail category 80 goes together with $m = 5$. For explanations about category 36*, see section 4.1.2.
71	$50 < \ell \leq 80$	all t			
63	$80 < \ell \leq 100$	all t			
56	$100 < \ell \leq 120$	all t			
56	$\ell > 120$	$t \leq 20$			
50	$120 < \ell \leq 200$	$t > 20$			
	$\ell > 200$	$20 < t \leq 30$			
45	$200 < \ell \leq 300$	$t > 30$			
	$\ell > 300$	$30 < t \leq 50$			
40	$\ell > 300$	$t > 50$			
As detail 1 in Table 8.5					
36*					

B.5 Load Carrying Welded Joints

As detail 1 in Table 8.5		Overlapped welded joints: 4) Fillet welded lap joint.	4) $\Delta\sigma$ in the main plate to be calculated on the basis of area shown in the sketch. 5) $\Delta\sigma$ to be calculated in the overlapping plates. Details 4) and 5): - Weld terminations more than 10 mm from plate edge. - Shear cracking in the weld should be checked using detail 8).	4) and 5) Welding quality level C could be specified. 5) For explanations about category 45*, see section 4.1.2.
45*		Overlapped: 5) Fillet welded lap joint.		
56*	$t_c < t$: —, $t_c \geq t$: —	Cover plates in beams and plate girders: 6) End zones of single or multiple welded cover plates, with or without transverse end weld.	6) If the cover plate is wider than the flange, a transverse end weld is needed. This weld should be carefully ground to remove undercut. The minimum length of the cover plate is 300 mm. For shorter attachments size effects see detail 1).	
50	$20 < t_c \leq 30$, $t \leq 20$			
45	$30 < t_c \leq 50$, $20 < t \leq 30$			
40	$t > 50$, $30 < t \leq 50$			
36	—, $t > 50$			
56		7) Cover plates in beams and plate girders. $5t_c$ is the minimum length of the reinforcement weld.	7) Transverse end weld ground flush. In addition, for $t_c > 20$ mm, front of plate at the end ground with a slope $< 1/4$.	
80 $m = 5$		8) Continuous fillet welds transmitting a shear flow, such as web to flange welds in plate girders. 9) Fillet welded lap joint.	8) $\Delta\tau$ to be calculated from the weld throat area. 9) $\Delta\tau$ to be calculated from the weld throat area considering the total length of the weld. Weld terminations more than 10 mm from the plate edge, see also 4) and 5) above.	

B. Fatigue Detail Tables with Commentary

Category	Detail	Description	Commentary
see EN 1994-2 (90, $m=8$)	Welded stud shear connectors: 10) For composite application	10) $\Delta\tau$ to be calculated from the nominal cross section of the stud.	EN 1994-2, Annex C.2, gives specific rules and fatigue strength for studs disposed horizontally and causing splitting forces in the direction of the slab thickness.
71	11) Tube socket joint with 80% full penetration butt welds.	11) Weld toe ground. $\Delta\sigma$ computed in tube.	Usually welding quality level B. Type of joint usually used in pipelines and pressure shafts, in order to avoid any perturbation in the flow of liquid in the tube.
40	12) Tube socket joint with fillet welds.	12) $\Delta\sigma$ computed in tube.	Welding quality level C could be specified. Type of joint usually used in chimneys.

B.6 HOLLOW SECTIONS (T ≤ 12.5 MM) (EN 1993-1-9, TABLE 8.6)

Detail category	Constructional detail	Description	Requirements	Commentary
71		1) Tube-plate joint, tubes flatted, butt weld (X-groove)	1) $\Delta\sigma$ computed in tube. Only valid for tube diameter less than 200 mm.	Alternative constructional detail can be developed.
71	$\alpha \leq 45°$	2) Tube-plate joint, tube slitted and welded to plate. Holes at end of slit.	2) $\Delta\sigma$ computed in tube. Shear cracking in the weld should be verified using Table 8.5 detail 8).	Crack may also start at end of weld near the hole. Alternative detailing can be developed.
63	$\alpha > 45°$			
71		Transverse butt welds: 3) Butt-welded end-to-end connections between circular structural hollow sections.	Details 3) and 4): Weld convexity ≤ 10% of weld width, with smooth transitions. Welded in flat position, inspected and found free from defects outside the tolerances of EN 1090. Classify 2 detail categories higher if $t > 8$ mm.	3) and 4) $\Delta\sigma$ shall include the stress concentration factor to allow for any thickness change and for eccentricity (due to concentricity, thickness change, out of roundness), see section 3.4.2.
56		4) Butt-welded end-to-end connections between rectangular structural hollow sections.		

B. FATIGUE DETAIL TABLES WITH COMMENTARY

Detail category	Constructional detail	Description	Requirements	Commentary
71		Welded attachments: 5) Circular or rectangular structural hollow section, fillet-welded to another section.	5) Non load-carrying welds. Width parallel to stress direction $\ell \leq 100$ mm. Other cases see Table 8.4.	
50		Welded splices: 6) Circular structural hollow sections, butt-welded end-to-end with an intermediate plate.	Details 6) and 7): Load-carrying welds. Welds inspected and found free from defects outside the tolerances of EN 1090. Classify 1 detail category higher if $t > 8$ mm.	Details 6) to 9) By analogy to cruciform joint, the misalignment of the tube wall should not exceed 15 % of the thickness of the intermediate plate.
45		7) Rectangular structural hollow sections, butt welded end-to-end with an intermediate plate.		
40		8) Circular structural hollow sections, fillet-welded end-to-end with an intermediate plate.	Details 8) and 9): Load-carrying welds. Wall thickness $t \leq 8$ mm.	
36		9) Rectangular structural hollow sections, fillet-welded end-to-end with an intermediate plate.		

B.7 LATTICE GIRDER NODE JOINTS (EN 1993-1-9, TABLE 8.7)

Detail category	Constructional detail	Requirements	Commentary
90 $m = 5$ $\dfrac{t_0}{t_i} \geq 2.0$	Gap joints: Detail 1): K and N joints, circular structural hollow sections: (1)	Details 1) and 2): - Separate assessments needed for the chords and the braces. - For intermediate values of the ratio t_0/t_i, interpolate linearly between detail categories. - Fillet welds permitted for braces with wall thickness $t \leq 8$ mm. - t_0 and $t_i \leq 8$ mm - $35° \leq \theta \leq 50°$ - $b_0/t_0 \cdot t_0/t_i \leq 25$ - $d_0/t_0 \cdot t_0/t_i \leq 25$ - $0.4 \leq b_i/b_0 \leq 1.0$ - $0.25 \leq d_i/d_0 \leq 1.0$	$\Delta\sigma$ computed in the diagonals and in the chord, with the effects of secondary bending according to Tables 4.1 or 4.2 of EN 1993-1-9 (k_1-factors). In total, three verifications to be carried out with the relevant detail category.
45 $m = 5$ $\dfrac{t_0}{t_i} = 1.0$			
71 $m = 5$ $\dfrac{t_0}{t_i} \geq 2.0$	Gap joints: Detail 2): K and N joints, rectangular structural hollow sections: (2)	- $b_0 \leq 200$ mm - $d_0 \leq 300$ mm - $0.5h_0 \leq e_{i/p} \leq 0.25h_0$ - $0.5d_0 \leq e_{i/p} \leq 0.25d_0$ - $e_{o/p} \leq 0.02b_0$ or $\leq 0.02d_0$ [$e_{o/p}$ is out-of-plane eccentricity] Detail 2): $0.5(b_0 - b_i) \leq g \leq 1.1(b_0 - b_i)$ and $g \geq 2t_0$	
36 $m = 5$ $\dfrac{t_0}{t_i} = 1.0$			

B. Fatigue Detail Tables with Commentary

71 $m = 5$	$\dfrac{t_0}{t_i} \geq 1.4$	Overlap joints: Detail 3): K joints, circular or rectangular structural hollow sections: ③	Details 3) and 4): - 30 % ≤ overlap ≤ 100 % - overlap = $(q/p) \times 100$ % - Separate assessments needed for the chords and the braces. - For intermediate values of the ratio t_0/t_i, interpolate linearly between detail categories. - Fillet welds permitted for braces with wall thickness $t \leq 8$ mm. - t_0 and $t_i \leq 8$ mm - $35° \leq \theta \leq 50°$ - $b_0/t_0 \cdot t_0/t_i \leq 25$ - $d_0/t_0 \cdot t_0/t_i \leq 25$ - $0.4 \leq b_i/b_0 \leq 1.0$ - $0.25 \leq d_i/d_0 \leq 1.0$ - $b_0 \leq 200$ mm - $d_0 \leq 300$ mm - $-0.5h_0 \leq e_{ip} \leq 0.25h_0$ - $-0.5d_0 \leq e_{ip} \leq 0.25d_0$ - $e_{o/p} \leq 0.02b_0$ or $\leq 0.02d_0$ ($e_{o/p}$ is out-of-plane eccentricity). Definition of p and q:
56 $m = 5$	$\dfrac{t_0}{t_i} = 1.0$		
71 $m = 5$	$\dfrac{t_0}{t_i} \geq 1.4$	Overlap joints: Detail 4): N joints, circular or rectangular structural hollow sections: ④	$\Delta\sigma$ computed in the diagonals and in the chord, with the effects of secondary bending according to Tables 4.1 or 4.2 of EN 1993-1-9 (k_1-factors). In total, three verifications to be carried out with the relevant detail category.
50 $m = 5$	$\dfrac{t_0}{t_i} = 1.0$		

B.8 ORTHOTROPIC DECKS – CLOSED STRINGERS (EN 1993-1-9, TABLE 8.8)

Detail category		Constructional detail	Description	Requirements	Comments
80	$t \leq 12$ mm		1) Continuous longitudinal stringer, with additional cutout in cross girder.	1) Assessment based on the direct stress range $\Delta\sigma$ in the longitudinal stringer.	See EN 1993-2, Annex C. See fatigue detail tables with commentary, Table B.13.
71	$t > 12$ mm				
80	$t \leq 12$ mm		2) Continuous longitudinal stringer, no additional cutout in cross girder.	2) Assessment based on the direct stress range $\Delta\sigma$ in the stringer.	
71	$t > 12$ mm				
36			3) Separate longitudinal stringer each side of the cross girder.	3) Assessment based on the direct stress range $\Delta\sigma$ in the stringer.	
71			4) Joint in rib, full penetration butt weld with steel backing plate.	3) Assessment based on the direct stress range $\Delta\sigma$ in the stringer.	

B. FATIGUE DETAIL TABLES WITH COMMENTARY

B.10 TOP FLANGE TO WEB JUNCTION OF RUNWAY BEAMS (EN 1993-1-9, TABLE 8.10)

Detail category	Constructional detail	Description	Requirements	Comments
160	①	1) Rolled I- or H-sections	1) Vertical compressive stress range $\Delta\sigma_{vert}$ in web due to wheel loads	$\Delta\sigma_{vert} \Leftrightarrow \Delta\sigma_{z,local}$
71	②	2) Full penetration tee-butt weld	2) Vertical compressive stress range $\Delta\sigma_{vert}$ in web due to wheel loads	Detail of weld:
36*	③	3) Partial penetration tee-butt welds, or effective full penetration tee-butt weld conforming with EN 1993-1-8	3) Stresses range $\Delta\sigma_{vert}$ in weld throat due to vertical compression from wheel loads	Welding quality level C could be specified. Insure good fit. Detail of weld:
36*	④	4) Fillet welds	4) Stresses range $\Delta\sigma_{vert}$ in weld throat due to vertical compression from wheel loads	Welding quality level C could be specified. Insure good fit. Detail of weld:

B.10 Top Flange to Web Junction of Runway Beams

71	⑤	5) T-section flange with full penetration tee-butt weld	5) Vertical compressive stress range $\Delta\sigma_{vert}$ in web due to wheel loads	Detail of weld:
36*	⑥	6) T-section flange with partial penetration tee-butt weld, or effective full penetration tee-butt weld conforming with EN 1993-1-8	6) Stresses range $\Delta\sigma_{vert}$ in weld throat due to vertical compression from wheel loads	Welding quality level C could be specified. Insure good fit. Detail of weld:
36*	⑦	7) T-section flange with fillet welds	7) Stresses range $\Delta\sigma_{vert}$ in weld throat due to vertical compression from wheel loads	Welding quality level C could be specified. Insure good fit. Detail of weld:

B. FATIGUE DETAIL TABLES WITH COMMENTARY

B.11 DETAIL CATEGORIES FOR USE WITH GEOMETRIC (HOT SPOT) STRESS METHOD (EN 1993-1-9, TABLE B.1)

Detail category	Constructional detail	Description	Requirements	Comments
112	①	1) Full penetration butt joint.	1) - All welds ground flush to plate surface parallel to direction of the arrow. - Weld run-on and run-off pieces to be used and subsequently removed, plate edges to be ground flush in direction of stress. - Welded from both sides, checked by NDT. For misalignment see NOTE 1.	
100	②	2) Full penetration butt joint.	2) - Weld not ground flush - Weld run-on and run-off pieces to be used and subsequently removed, plate edges to be ground flush in direction of stress. - Welded from both sides. For misalignment see NOTE 1.	Should consider thickness correction: $K_s = (25/t)^{0.2} < 1.0$
100	③	3) Cruciform joint with full penetration K-butt welds.	3) - Weld toe angle ≤ 60°. - For misalignment see NOTE 1.	Should consider thickness correction: $K_s = (25/t)^{0.2} < 1.0$ Currently not calibrated for hollow section joints. See specific S-N curves given in (CIDECT, 2000) or (IIW, 2000).
100	④	4) Non-load carrying fillet welds.	4) - Weld toe angle ≤ 60°. - See also NOTE 2.	Should consider thickness correction: $K_s = (25/t)^{0.2} < 1.0$ See comment under detail 3)

B.11 DETAIL CATEGORIES FOR USE WITH GEOMETRIC STRESS METHOD

100	(5)	5) Bracket ends, ends of longitudinal stiffeners.	5) - Weld toe angle ≤ 60°. - See also NOTE 2.
100	(6)	6) Cover plate ends and similar joints.	6) - Weld toe angle ≤ 60°. - See also NOTE 2.
90	(7)	7) Cruciform joints with load-carrying fillet welds.	7) - Weld toe angle ≤ 60°. - For misalignment see NOTE 1. - See also NOTE 2. Should consider thickness correction: $$K_s = (25/t)^{0.2} < 1.0$$ See comment under detail 3)

NOTE 1 Table B.1 does not cover effects of misalignment. They have to be considered explicitly in determination of stress.
NOTE 2 Table B.1 does not cover fatigue initiation from the root followed by propagation through the throat.

287

B. FATIGUE DETAIL TABLES WITH COMMENTARY

B.12 TENSION COMPONENTS

Detail category [N/mm²] for exposure class 3 or 4	Constructional detail	Component description	Comments
***		Tension rod (bar) system	Group of tension components A) Single solid round cross section connected to end terminations by threads
105		Prestressing bar	
***		Spiral, circular strand rope. Typical diameter range of 5 mm to 160 mm.	Group of tension components B) Ropes composed of wires or stands (in spiral) which are anchored in sockets or other end terminations
150		Fully locked circular coil rope with metal or resin socketing. Typical diameter range of 20 mm to 180 mm	

B.12 TENSION COMPONENTS

150	Strands with metal or resin socketing	
160	Parallel wire strand (PWS) with epoxy socketing	Group of tension components C) Products composed of parallel wires or parallel strands needing individual or collective anchoring and appropriate protection
160	Bundle of parallel strands (seven wire prestressing)	
160	Bundle of parallel wires	
***	Multiple bars	

*** to be determined by tests. Specific requirements for fatigue testing of wire, strands, bars and complete tension components are given in EN 1993-1-11, Annex A.

B. FATIGUE DETAIL TABLES WITH COMMENTARY

B.13 REVIEW OF ORTHOTROPIC DECKS DETAILS AND STRUCTURAL ANALYSIS

Constructional detail	Structural analysis	Relevant stresses and categories $\Delta\sigma_c$ [N/mm²]
Continuous trough to crossbeam, with cope hole, trough failures[1]	Trough as continuous beam on elastic supports (simplified analysis)[3] t = thickness of crossbeam	*Nominal stress* 1a) In the trough web at the crossbeam, along vertical weld $\gamma_{Mf} = 1.0$ $\Delta\sigma_C\ (t \leq 12\text{ mm}) = 80$ $\Delta\sigma_C\ (t > 12\text{ mm}) = 71$ *Nominal stress* 1b) In the trough web at the crossbeam, around lower end of connection[2] $\gamma_{Mf} = 1.0$ $\Delta\sigma_C\ (t \leq 12\text{ mm}) = 80$ $\Delta\sigma_C\ (t > 12\text{ mm}) = 71$
Trough to crossbeam joint, continuous trough, close fit, crossbeam web or trough failures	Trough as continuous beam on elastic supports (simplified analysis)[3] [3] instead of a simplified analysis, a full model of the members crossing (crossbeam and troughs) can be made to evaluate properly out-of-plane behaviour of crossbeam web and resulting stresses.	*Nominal stress* 2a), 2b) In the trough $\gamma_{Mf} = 1.0$ $\Delta\sigma_C\ (t \leq 12\text{ mm}) = 80$ $\Delta\sigma_C\ (t > 12\text{ mm}) = 71$ [1] critical section in web of crossbeam due to cut-outs. The failure location depends upon geometric parameters and thicknesses, for example the size, shape of the cope hole. The cope hole geometries showed are indicative, i.e. cracking case b can occur at a cope hole presented under a and vice-versa. [2] For cracks at the weld toe along the vertical weld in the joint with a cope hole, a geometric stress approach seems more appropriate. However, not enough data is available to define a detail category using it. For the time being, the nominal stress approach is kept (Kolstein, 2007).

Constructional detail	Structural analysis	Relevant stresses and categories $\Delta\sigma_c$ [N/mm^2]
Trough to crossbeam joint, discontinuous trough, trough failures	At the soffit of the trough at the crossbeam	*Nominal stress* 3a), 3b) In the trough Failure from the root $\gamma_{Mf} = 1.15$ $\Delta\sigma_C = 36$ Failure from the toe $\gamma_{Mf} = 1.0$ $\Delta\sigma_C = 80$
Trough splice joint, full penetration butt weld	Trough as continous beam on elastic supports	*Nominal stress* 4) on steel backing, failure in the trough $\gamma_{Mf} = 1.0$ $\Delta\sigma_C = 71$ 5) without backing, failure in the trough $\gamma_{Mf} = 1.0$ Category in function of butt weld type: $\Delta\sigma_C = 112$ (as detail 1, 2, 4 Table 8.3) $\Delta\sigma_C = 90$ (as detail 5, 7 Table 8.3) $\Delta\sigma_C = 80$ (as detail 9, 11 Table 8.3)

Constructional detail	Structural analysis	Relevant stresses and categories $\Delta\sigma_c$ [N/mm^2]
Continuous trough to crossbeam joint with cope hole, crossbeam web failures[4]	Crossbeam as Vierendeel beam model in bending, see EN 1993-2, 9.4.2[5]	6a) At the free edge of the cope hole $\gamma_{Mf} = 1.15$ *Nominal stress* $\Delta\sigma_c = 71$ *Geometric stress* $\Delta\sigma_c = 112$
	Trough as continuous beam on elastic supports (simplified analysis)[6]	*Nominal stress* 6b) Inside the cope hole at the web weld toe $\gamma_{Mf} = 1.0$ $\Delta\sigma_c = 71$
Trough to deck plate joint partial penetration weld	Deck plate as continous beam on elastic supports (the troughs), transverse bending t_t = thickness of trough	*Nominal stress* 7) In the deck plate at the trough $\gamma_{Mf} = 1.0$ $\Delta\sigma_c = 71$ Partial penetration weld with $a \geq t_t$ and gap ≤ 1 mm (Kolstein, 2007). If not satisfied, see detail 8)

[4] critical section in web of crossbeam due to cut-outs. The failure location depends upon geometric parameters and thicknesses, for example the size and shape of the cope hole. The cope holes geometry showed are indicative, i.e. cracking cases b can occur at a cope hole presented under a and vice-versa.
[5] In addition to the stresses from the crossbeam in bending, the crossbeam web is under some imposed out-of-plane rotation range, which is neglected in the structural model.
[6] instead of a simplified analysis, a full model of the members crossing (crossbeam and troughs) can be made to evaluate properly out-of-plane behaviour of crossbeam web and resulting stresses.

Constructional detail	Structural analysis	Relevant stresses and categories $\Delta\sigma_c$ [N/mm^2]
Trough to deck plate joint	Deck plate beam on elastic supports, trough web plate in transverse bending t_t = thickness of trough	*Nominal stress* 8) In the trough web at the deck plate $\gamma_{Mf} = 1.0$ $\Delta\sigma_c = 50$ Fillet or partial penetration weld with $a < t_t$ or gap > 1 mm
Trough to deck plate and crossbeam joint[7]	Trough as continuous beam on elastic supports (simplified analysis) [7] details and categories not explicitly given in EN 1993-1-9 tables for orthotropic decks (i.e. cases of longitudinal welds and transverse attachments and butt welds), but which must be checked.	*Nominal stress* 9a) At the weld root in the deckplate, $\gamma_{Mf} = 1.0$ Category in function of butt weld type: $\Delta\sigma_c = 112$ (as detail 1, 2, 4 Table 8.3) $\Delta\sigma_c = 90$ (as detail 5, 7 Table 8.3) $\Delta\sigma_c = 80$ (as detail 9, 11 Table 8.3)
Crossbeam or longitudinal web to deck plate joint[7] See drawing above		9b) In the deck plate at the location of the trough, $\gamma_{Mf} = 1.0$ $\Delta\sigma_c = 112$ (detail 3, table 8.2) $\Delta\sigma_c = 100$ (detail 5,6,7 table 8.2)
Butt joints in the deck plate[7] See drawing above		9c) In the longitudinal web at the deck plate, $\gamma_{Mf} = 1.0$ $\Delta\sigma_c$ ($t \leq 12$ mm) = 80 $\Delta\sigma_c$ ($t > 12$ mm) = 71

C. MAXIMUM PERMISSIBLE THICKNESSES TABLES

Steel grade	Sub-grade	Charpy Energy CVN at T [°C]	J_{min}	Reference temperature T_{Ed} [°C]																		
				$\sigma_{Ed} = 0.75 f_y(t)$							$\sigma_{Ed} = 0.50 f_y(t)$							$\sigma_{Ed} = 0.25 f_y(t)$						
				10	0	-10	-20	-30	-40	-50	10	0	-10	-20	-30	-40	-50	10	0	-10	-20	-30	-40	-50
S460	Q	-20	30	70	60	50	40	30	25	20	110	95	75	65	55	45	35	175	155	130	115	95	80	70
	M,N	-20	40	90	70	60	50	40	30	25	130	110	95	75	65	55	45	200	175	155	130	115	95	80
	QL	-40	30	105	90	70	60	50	40	30	155	130	110	95	75	65	55	200	200	175	155	130	115	95
	ML,NL	-50	27	125	105	90	70	60	50	40	180	155	130	110	95	75	65	200	200	200	175	155	130	115
	QL1	-60	30	150	125	105	90	70	60	50	200	180	155	130	110	95	75	215	200	200	200	175	155	130
S690	Q	0	40	40	30	25	20	15	10	10	65	55	45	35	30	20	20	120	100	85	75	60	50	45
	Q	-20	30	50	40	30	25	20	15	10	80	65	55	45	35	30	20	140	120	100	85	75	60	50
	QL	-20	40	60	50	40	30	25	20	15	95	80	65	55	45	35	30	165	140	120	100	85	75	60
	QL	-40	30	75	60	50	40	30	25	20	115	95	80	65	55	45	35	190	165	140	120	100	85	75
	QL1	-40	40	90	75	60	50	40	30	25	135	115	95	80	65	55	45	200	190	165	140	120	100	85
	QL1	-60	30	110	90	75	60	50	40	30	160	135	115	95	80	65	55	200	200	190	165	140	120	100

C.2 MAXIMUM PERMISSIBLE VALUES OF ELEMENT THICKNESS T IN MM (EN 1993-1-12, TABLE 4)

Steel grade	Sub-grade	Charpy Energy CVN at T [°C]	J_{min}	Reference temperature T_{Ed} [°C]																		
				$\sigma_{Ed} = 0.75 f_y(t)$							$\sigma_{Ed} = 0.50 f_y(t)$							$\sigma_{Ed} = 0.25 f_y(t)$						
				10	0	-10	-20	-30	-40	-50	10	0	-10	-20	-30	-40	-50	10	0	-10	-20	-30	-40	-50
EN 10025-6																								
S500	Q	0	40	55	45	35	30	20	15	15	85	70	60	50	40	35	25	145	125	105	90	80	65	55
	Q	-20	30	65	55	45	35	30	20	15	105	85	70	60	50	40	35	170	145	125	105	90	80	65
	QL	-20	40	80	65	55	45	35	30	20	125	105	85	70	60	50	40	195	170	145	125	105	90	80
	QL	-40	30	100	80	65	55	45	35	30	145	125	105	85	70	60	50	200	195	170	145	125	105	90
	QL1	-40	40	120	100	80	65	55	45	35	170	145	125	105	85	70	60	200	200	195	170	145	125	105
	QL1	-60	30	140	120	100	80	65	55	45	200	170	145	125	105	85	70	205	200	200	195	170	145	125
S550	Q	0	40	50	40	30	25	20	15	10	80	65	55	45	35	25	25	140	120	100	85	75	60	50
	Q	-20	30	60	50	40	30	25	20	15	95	80	65	55	45	35	30	160	140	120	100	85	75	60
	QL	-20	40	75	60	50	40	30	25	20	115	95	80	65	55	45	35	185	160	140	120	100	85	75
	QL	-40	30	90	75	60	50	40	30	25	135	115	95	80	65	55	45	200	185	160	140	120	100	85
	QL1	-40	40	110	90	75	60	50	40	30	160	135	115	95	80	65	55	200	200	185	160	140	120	100
	QL1	-60	30	120	110	90	75	60	50	40	185	160	135	115	95	80	65	200	200	200	185	160	140	120
S620	Q	0	40	45	35	25	20	15	15	10	70	60	50	40	30	25	20	130	110	95	80	65	55	45
	Q	-20	30	55	45	35	25	20	15	15	85	70	60	50	40	30	25	150	130	110	95	80	65	55
	QL	-20	40	65	55	45	35	25	20	15	105	85	70	60	50	40	30	175	150	130	110	95	80	65

C.2 MAXIMUM PERMISSIBLE VALUES OF ELEMENT THICKNESS T IN MM

Steel grade	Sub-grade	Charpy Energy CVN at T [°C]	J_{min}	Reference temperature T_{Ed} [°C]																			
				10	0	-10	-20	-30	-40	-50	10	0	-10	-20	-30	-40	-50	10	0	-10	-20	-30	-40	-50
				$\sigma_{Ed} = 0.75\, f_y(t)$							$\sigma_{Ed} = 0.50\, f_y(t)$							$\sigma_{Ed} = 0.25\, f_y(t)$						
S620	QL	-40	30	80	65	55	45	35	25	20	125	105	85	70	60	50	40	200	175	150	130	110	95	80
	QL1	-40	40	100	80	65	55	45	35	25	145	125	105	85	70	60	50	200	200	175	150	130	110	95
	QL1	-60	30	120	100	80	65	55	45	35	170	145	125	105	85	70	60	200	200	200	175	150	130	110
S690	Q	0	40	40	30	25	20	15	10	10	65	55	45	35	30	20	20	120	100	85	75	60	50	45
	Q	-20	30	50	40	30	25	20	15	10	80	65	55	45	35	30	20	140	120	100	85	75	60	50
	QL	-20	40	60	50	40	30	25	20	15	95	80	65	55	45	35	30	165	140	120	100	85	75	60
	QL	-40	30	75	60	50	40	30	25	20	115	95	80	65	55	45	35	190	165	140	120	100	85	75
	QL1	-40	40	90	75	60	50	40	30	25	135	115	95	80	65	55	45	200	190	165	140	120	100	85
	QL1	-60	30	110	90	75	60	50	40	30	160	135	115	95	80	65	55	200	200	190	165	140	120	100
EN 10149-2																								
S500	MC	-20	40	80	65	55	45	35	30	20	125	105	85	70	60	50	40	195	170	145	125	105	90	80
S550	MC	-20	40	75	60	50	40	30	25	20	115	95	80	65	55	45	35	185	160	140	120	100	85	75
S600	MC	-20	40	70	55	45	35	30	20	15	105	90	75	60	50	40	35	180	155	130	110	95	80	70
S650	MC	-20	40	65	50	40	30	25	20	15	100	85	70	55	45	35	30	170	145	125	105	90	75	65
S700	MC	-20	40	60	45	35	30	25	20	15	95	80	65	50	45	35	30	165	140	120	100	85	70	60